Microparticles in Stokes Flows 2011: Symposium in Honor of Francois Feuillebois' 65th Birthday

Journal of Physics: Conference Series Volume 392

Warsaw, Poland
21-24 August 2011

Editors:

Maria L Ekiel-Jeżewska

ISBN: 978-1-62276-856-1
ISSN: 1742-6588

Microparticles in Stokes Flows 2011: Symposium in Honor of Francois Feuillebois' 65th Birthday

Journal of Physics: Conference Series Volume 392

Warsaw, Poland
21-24 August 2011

ISBN: 978-1-62276-856-1
ISSN: 1742-6588

IOP Institute of Physics

Printed from e-media with permission by:

Curran Associates, Inc.
57 Morehouse Lane
Red Hook, NY 12571

Some format issues inherent in the e-media version may also appear in this print version.

Printed by Curran Associates, Inc. (2013)

For permission requests, please contact the Institute of Physics
at the address below.

Institute of Physics
Dirac House, Temple Back
Bristol BS1 6BE UK

Phone: 44 1 17 929 7481
Fax: 44 1 17 920 0979

techtracking@iop.org

Additional copies of this publication are available from:

Curran Associates, Inc.
57 Morehouse Lane
Red Hook, NY 12571 USA
Phone: 845-758-0400
Fax: 845-758-2634
Email: curran@proceedings.com
Web: www.proceedings.com

TABLE OF CONTENTS

Preface

Microparticles in Stokes Flows – Symposium in Honor of François Feuillebois' 65th Birthday was held from 21–24 August 2011 in Warsaw, Poland. There were 43 participants from Australia, Brazil, France, Germany, Ireland, Israel, Japan, the Netherlands, Poland, Russia, Tunisia, United Kingdom and the USA. They presented 5 invited lectures, 25 contributed talks and 9 posters. The abstracts can be found at microparticles2011.ippt.pan.pl. The subjects were divided into three parts and 13 sessions, listed below.

Part I. *A variety of particles*. Sessions: sedimentation at non-Brownian time scale, Brownian and hydrodynamic interactions of non-spherical particles, active particles, permeable particles, deformable particles and biosystems.

Part II. *Particles and interfaces*. Sessions: confined multiparticle systems, hydrodynamic interactions in confined systems, single particle under confinement, particle-interface interactions, slip at interfaces, slip in suspensions.

Part III. *Stokes flows and beyond*. Sessions: inertial effects, texture and self-assembly.

At the end of each session, presentations were followed by extensive discussions (two thirds of the presentation time). Also, two round-table general discussions took place, *Methods of solving the Stokes equations - which when?* and *Challenges in microhydrodynamics*.

This volume contains both original contributions and reviews. There were at least two referees per paper: one participating in the conference, and the other not. I would like to thank all the referees for their work, the Polish Academy of Sciences, Centre national de la recherche scientifique and the Institute of Fundamental Technological Research, Polish Academy of Sciences for supporting in part the conference, Anna Myłyk, Agnieszka Słowicka and Krzysztof Zembrzycki for the weeks spent organizing the symposium, and all the conference participants for their presentations and discussions of the highest scientific quality.

Journal of Physics: Conference Series **392** (2012) 011001 doi:10.1088/1742-6596/392/1/011001

Happy Birthday, François Feuillebois!

This volume of *Journal of Physics: Conference Series* is dedicated to our friend and colleague François Feuillebois. The symposium *Microparticles in Stokes Flows* was organized to celebrate his 65th birthday. François presented a special lecture, entitled *Influence of wall slip in dilute suspensions* [1], which was followed by a birthday dinner, with an after-dinner speech given by John Hinch. The symposium participants, and those who could not join the meeting, all wish François all the best for his 65th birthday, and thank him for sharing with us his knowledge, and motivating us to solve interesting problems related to motion of particles in fluid flows.

An illustration of François's scientific interests was shown on the conference poster.

The conference participants could participate in a contest, by submitting a written answer to the question "What system is shown on the conference poster (equations, boundary conditions)?". Francois was in charge of evaluating the answers to this quiz, and providing the explanation [3]. Transworld Research Network kindly provided the rewards: copies of the book co-edited by Francois [2]. The winners were Maciej Lisicki and Eric Climent. A third copy of this book was awarded to John Hinch, who (according to majority of the participants) had asked the most insightful questions during the conference.

Maria L Ekiel-Jeżewska (Editor)
Institute of Fundamental Technological Research, Polish Academy of Sciences, Pawińskiego 5B, 02-106 Warsaw, Poland
Email: mekiel@ippt.pan.pl

References

[1] Feuillebois F, Ghalya N and Sellier A 2012 Influence of wall slip in dilute suspensions *J. Phys.: Conf. Ser.* **392** 012012.
[2] Feuillebois F and Sellier A, eds., 2009 *Theoretical Methods for Micro Scale Viscous Flows* (Transworld Research Network, Kerala).
[3] Feuillebois F 2012 The quiz *J. Phys.: Conf. Ser.* **392** 011002.

Journal of Physics: Conference Series **392** (2012) 011002 doi:10.1088/1742-6596/392/1/011002

The quiz

François Feuillebois
LIMSI-CNRS, UPR 3251, B.P. 133, 91403 Orsay cédex, France
E-mail: francois.feuillebois@limsi.fr

Abstract. The flow field represented in the following figure was shown on the cover of the Symposium book of abstracts. Its interpretation was proposed to the participants as a quiz. The answer to the quiz is given here.

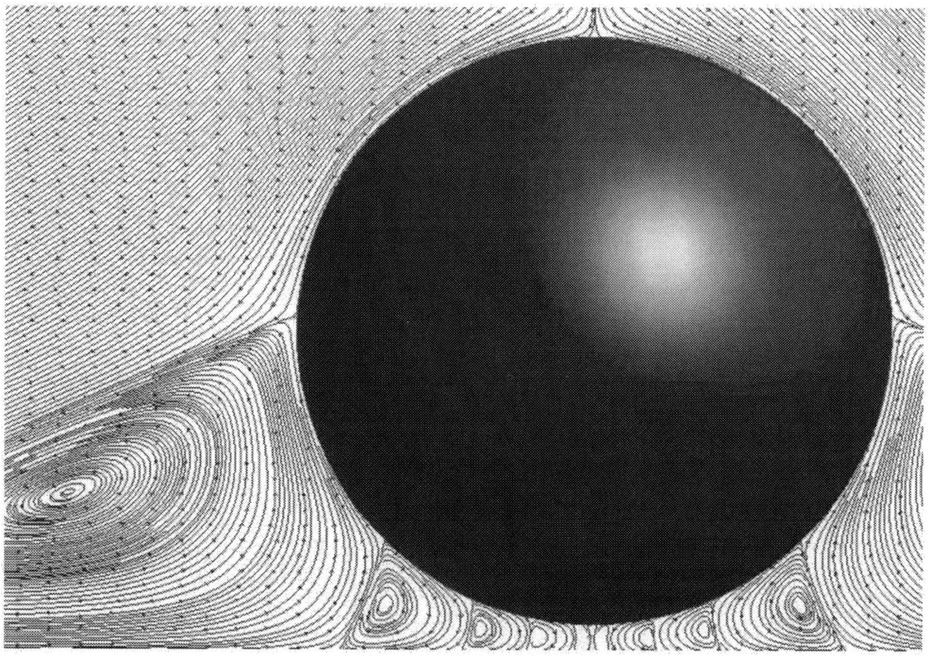

The proposed figure is another view of the flow field plotted in [1], Fig. 11 (right)[1]. This is a creeping flow, where the fluid velocity $\mathbf{v}$ and pressure p are solutions of Stokes equations:

$$\boldsymbol{\nabla} \cdot \mathbf{v} = 0\,, \qquad \boldsymbol{\nabla} p = \mu \nabla^2 \mathbf{v}\,,$$

in which μ denotes the fluid viscosity. The fluid flows around a fixed spherical particle with radius a and centre at a distance $1.1\,a$ from a solid plane wall (located at the bottom line of the figure). The no-slip boundary condition $\mathbf{v} = 0$ applies on the particle and wall.

Consider cylindrical coordinates ρ, z, with z representing the distance from the wall and ρ the distance from the symmetry axis. The particle is embedded in an unperturbed axisymmetric

[1] Reprinted with permission ©2005, American Institute of Physics.

stagnation point flow, for which the velocity varies as a polynomial of degree 3 in the coordinates. More specifically, the velocity components and pressure for this flow field satisfying the Stokes equations are:

$$u_\rho = S_3 \rho z^2, \qquad u_z = -2S_3 z^3, \qquad p = -3S_3\mu(-2z^2 - \rho^2),$$

where S_3 is a constant.

The axisymmetric flow field perturbed by the particle was calculated with high accuracy in bispherical coordinates. It presents interesting toroidal recirculation regions, or toroidal vortices, a section of which is visible here in a meridian plane (The outmost largest vortex is not fully shown in this figure). Note that the classical stagnation point flow (also considered in [1]), for which the velocity varies as a polynomial of degree 2 in the coordinates, creates similar but not so wide vortices.

The physical origin of these vortices is quite different from those due to fluid inertia. Indeed, there is no wake in creeping flow. The physical interpretation is that a very viscous fluid "avoids" the regions where it would dissipate too much energy. Calculations show that these vortices become smaller and more numerous as the sphere is closer to the wall. In the limit of a vanishing gap (a case which, however, cannot be calculated with the bispherical coordinates method), it is anticipated that there might be an infinite sequence of nested toroidal vortices, alike the infinite set of nested vortices found by Moffatt [2] for the two-dimensional flow field in a wedge.

References

[1] Pasol L, Chaoui M, Yahiaoui S and Feuillebois F 2005 *Physics of Fluids* **17** 073602
[2] Moffatt H K 1964 *J. Fluid Mech* **18** 1–18

Interview with François Feuillebois

Maria L. Ekiel-Jeżewska: François, you have been investigating hydrodynamics of microparticles in viscous fluids for about forty years. How did you get motivated?

François Feuillebois: Under the guidance of André Lasek, I started as a PhD student to work in the field of suspensions. In 1972, he was back from DAMTP, Cambridge, UK, and asked me to read the now celebrated paper by Batchelor on sedimentation [1]. I then engaged to work on sedimentation and various other problems for suspensions. As an example, after staying as a post-doc in the group of Batchelor at Cambridge, I derived theoretical results for inhomogeneous sedimentation [2].

Was the personal contact with Batchelor essential for your scientific development?

Indeed, when George Batchelor visited the CNRS Aérothermique lab in Meudon before I went to Cambridge, I was starting on the sedimentation problem and he gave me some precious hints. When at DAMTP, Cambridge, he offered me to work on the problem of inhomogeneous sedimentation at the top of a settling cloud of particles, a difficult problem indeed. I then also benefited from useful discussions with him and his collaborators, namely John Hinch, David Jeffrey, John Rallison.

How has it later happened that you have extensively collaborated with Polish scientists, and visited Poland so often?

Thanks to André Lasek who was originally from Poland, I also knew about the Biennial Fluid Mechanics Conferences organised by Professor Fiszdon in Poland. I attended the ones in Białowieża (1975), Błażejewko (1979), Jachranka (1981), Sobieszewo (1985), Mrągowo (1987), Kozubnik (1989) and then in Warsaw (1992). During these symposia, I had the occasion to meet various colleagues, in particular from Poland and Bulgaria, with whom I continued thereafter an extensive collaboration. The present symposium is following a long fruitful collaboration with IPPT PAN [3], in particular with yourself, and I am grateful for this collaboration.

Why did you attend so many conferences in Poland?

At that time, these conferences were one of the very few opportunities for scientists of eastern Europe to meet those of western Europe, and conversely. Apart from the high level in fluid mechanics, there was a very friendly and informal atmosphere when people realised that, in some unknown little place somewhere in Poland, the only possibilities were to discuss and then enjoy come-together evenings.

What happened later with the tradition of the Biennial Fluid Mechanics Conferences?

The Fiszdon's conferences transformed into the Euromech Fluid Mechanics conferences when frontiers of eastern countries opened. I also attended the 1st Euromech Fluid Mechanics conference in Cambridge in 1991 and the 2nd one in Warsaw in 1994.

Although Professor Fiszdon was known in the international fluid community as the organiser of the Biennial Fluid Mechanics Conferences [4], but he considered educating a few people as the only own achievement [5]. What has been your contribution to education of next generations?

In the late nineties, when I visited Professor Fiszdon at home after his accident, the main question that he asked me was whether I had some PhD students. When I told him I had several at a time, he told me something like: "Oh yes, this is what I liked as well". Indeed, for Professor Fiszdon, supervising PhD students was the achievement. Today, I am happy to have been a guide for 25 PhD students who had themselves afterwards professional achievements. Also, when supervising them, they continuously provided me with an impulse to go further.

Could you tell young researchers how to succeed? How had you got motivated to persist in working on research problems, which you finally solved?

An interesting theoretical problem is one which may first appear as a paradox. In 1972, I wondered why do two spheres settle faster than one and many dispersed spheres in a container settle slower. But one is not always lucky in finding such a "paradox". In any case, the domain of fluid mechanics at micro-scales is rich with applications, in the meteorology of clouds, in analytical chemistry, etc. and discussions with colleagues familiar with such applications have been for me a continuing source of inspiration. I think that advances in difficult problems (typically like sedimentation, which is not fully solved today, in despite of recent advances published by colleagues in the present volume) deserve techniques which are novel for fluid mechanics, even though they may be well known in other scientific domains. For instance, the calculations in Ref. [2] were done after I learned some basis of the theory of distributions. Various experimental papers were connected to "novel" appropriate techniques, in collaboration with colleagues who are specialists of them; for instance, the measurement of the average velocity of sedimentation using X-ray absorption [6], the measurement of hydrodynamic interactions between spheres at contact using interferometry, one rough moving sphere being a reflector, see the paper [7] (following a collaboration with D. Joyeux at Institute of Optics, Orsay, in 1987-1991). More recently, the Aris-Taylor dispersion with a slip wall (see my paper herein) was solved for interpreting results of the novel Double-Focus Fluorescence Cross-Correlation technique to measure the slip length [8].

Well, do you think that progress is based on linking theory to experiments, and on collaboration with scientists of a different experience?

I realised that there are two basic approaches in fundamental science. One is advancing in theory and trying afterwards to set up an experiment which can verify it. Another one is to find with a simple experiment a new phenomenon and thereafter try to model it. Both approaches have their interest and both are used by colleagues. I tend to use the first one, favouring theory and collaborating with colleagues, in particular for experiments.

In your opinion, what are nowadays the fundamental open problems in microhydrodynamics, and how to search for solving them?

Microhydrodynamics is nowadays a rapidly evolving field with the new devices in microfluidics, etc. So some ideas that we have today may soon become obsolete. We can consider for instance micro-devices in which there are not only straight pipes but also a number of connections. For these sizes, low Reynolds number analysis may be possible, not only at first order (Stokes flow) but also at second order (with some fluid inertia), perhaps considering also slip on walls. The motion of fluid-fluid interfaces in such devices is also of importance. Keeping aware of the evolution in this field is certainly a continuing source of inspiration.

[1] Batchelor K G 1972 Sedimentation in a dilute dispersion of spheres *J. Fluid Mech.* **52** 245–68
[2] Feuillebois F 1984 Sedimentation in a dispersion with vertical inhomogeneities *J. Fluid Mech.* **139** 145–71
[3] Feuillebois F and Ekiel-Jeżewska M L 2010 Suspensions de particules et interactions hydrodynamiques dans un fluide visqueux *Annales, Centre Scientifique de l'Académie Polonaise de Sciences* **12** 44–61 (http://www.academie-polonaise.org/index.php?option=com_content&view=category&layout=blog&id=60&Itemid=66&lang=fr).
[4] Zachara A, Kowalewski T A and Łucjanek W 2005 Władysław Fiszdon (1912 2004), Obituary *Arch. Mech.* **57** 3–6; see also: fluid.ippt.pan.pl/fiszdon
[5] Herczyński R, Interview with Władysław Fiszdon, *IPPT Archives*, unpublished.
[6] Bruneau D, Anthore R, Feuillebois F, Auvray X and Petipas C 1990 Measurement of the velocity of sedimentation in a dilute polydisperse suspension of spheres *J. Fluid Mech.* **221** 577–96
[7] Ekiel-Jeżewska M L, Feuillebois F, Lecoq N, Masmoudi K, Anthore R, Bostel F and Wajnryb E 1999 Hydrodynamic interactions between two spheres at contact. *Phys. Rev. E* **59** 3182–91
[8] Vinogradova O I, Koynov K, Best A, and Feuillebois F 2009 Direct measurements of hydrophobic slippage using double-focus fluorescence cross-correlation *Phys. Rev. Lett.* **102** 118302

All papers published in this volume of *Journal of Physics: Conference Series* have been peer reviewed through processes administered by the proceedings Editors. Reviews were conducted by expert referees to the professional and scientific standards expected of a proceedings journal published by IOP Publishing.

Collective motion in ferrofluids

B. U. Felderhof

Institut für Theoretische Physik A RWTH Aachen University Templergraben 55 52056 Aachen Germany

E-mail: `ufelder@physik.rwth-aachen.de`

Abstract. Collective motion in a ferrofluid is discussed on the basis of ferrohydrodynamics. Simple examples elucidate the effect of a collection of spinning suspended particles on Stokes flow. Spin viscosity emerges as a crucial transport coefficient. In the case of a uniform rotating magnetic field applied to a ferrofluid in a cylinder or a spherical cavity the collective fluid motion is driven by the curl of the torque density, which differs from zero only in a thin boundary layer with thickness proportional to the square root of the spin viscosity.

1. Introduction

Ferrofluids find use in many technical applications[1]. Recent work shows that effective pumping of a ferrofluid by the application of a running magnetic wave is possible[2]. No doubt, this will lead to interesting novel applications. The corresponding theory is being developed, and gives rise to new questions and problems.

Ferrofluids have the important property that one can act on them with a magnetic field, since that couples to the magnetic dipole moments of suspended particles. The theoretical description is based on an extension of the Navier-Stokes equations, with account of the magnetic force and torque density acting on the fluid. Collective motions of the fluid can be generated by judicious application of a field which varies in space and time.

The equations of motion are complicated, since they couple the flow velocity and the rotational velocity of spinning particles to magnetic field and magnetization in a nonlinear way. Magnetic field and magnetization satisfy Maxwell's equations of magnetostatics, and this gives rise to long range effects, so that confinement and geometry crucially affect fluid behavior. The relaxation of magnetization to its thermal equilibrium value is described by a phenomenological relaxation equation, which involves a nonlinear coupling to flow velocity and spin velocity.

The hydrodynamic stress tensor of a ferrofluid is not symmetric[1], and involves the transport coefficients vortex viscosity ζ and spin viscosity η', besides the shear viscosity η. The vortex viscosity can be estimated fairly well for a dilute suspension as $\zeta = \frac{3}{2}\phi\eta$, where ϕ is the volume fraction of suspended particles. The spin viscosity is more problematic. We have suggested elsewhere[3], that it can possibly be determined in computer simulation from its effect in simple flow situations.

In the following we discuss first how a torque density can generate flow in a viscous incompressible fluid described by the Stokes equations. Subsequently we recall the equations of ferrohydrodynamics, and review recent work on the rotation of a ferrofluid, confined in a cylinder or a sphere, caused by the application of a uniform rotating magnetic field.

2. Torque density acting on a viscous incompressible fluid

In this section we consider a viscous incompressible fluid and study the flow caused by an applied torque density in some simple situations. The torque density may be caused by a rotating magnetic field acting on magnetic particles with permanent magnetic dipole moment. For the time being we neglect the interaction between dipole moments. We consider only phenomena on a slow timescale and a small length scale, so that we may use the equations of low Reynolds number hydrodynamics[4]. Thus we consider the steady-state Stokes equations

$$\eta\nabla^2\boldsymbol{v} - \nabla p = -\boldsymbol{F}, \qquad \nabla\cdot\boldsymbol{v} = 0, \tag{1}$$

where $\boldsymbol{v}(\boldsymbol{r})$ is the flow velocity, $p(\boldsymbol{r})$ is the pressure, and $\boldsymbol{F}(\boldsymbol{r})$ is the force density acting on the fluid via suspended particles. It is useful to express the force density, like the charge density in electrostatics, in terms of a multipole expansion about the particle centers[5]

$$\boldsymbol{F}(\boldsymbol{r}) = \boldsymbol{F}_1(\boldsymbol{r}) - \nabla\cdot\boldsymbol{F}_2^s(\boldsymbol{r}) + \frac{1}{2}\nabla\times\boldsymbol{T}(\boldsymbol{r}) + ..., \tag{2}$$

where $\boldsymbol{F}_1(\boldsymbol{r})$ is the monopole force density, $\boldsymbol{F}_2^s(\boldsymbol{r})$ is the symmetric force dipole density, and $\boldsymbol{T}(\boldsymbol{r})$ is the torque density, which may also be regarded as the antisymmetric force dipole density. In the absence of applied forces the monopole force density vanishes. The symmetric force dipole density is induced by local velocity gradients, and leads to an effective viscosity. Higher order force multipole densities will be neglected. Thus we consider instead of Eq. (2.1) the equations

$$\eta\nabla^2\boldsymbol{v} - \nabla p = -\frac{1}{2}\nabla\times\boldsymbol{T}, \qquad \nabla\cdot\boldsymbol{v} = 0, \tag{3}$$

where now η is the viscosity of the suspension, rather than that of the pure fluid. Due to the presence of the curl operator, regions of uniform torque density cause no flow, and only the spatial variation of the torque density matters.

For a uniform torque density $\boldsymbol{T}(\boldsymbol{r}) = T\boldsymbol{e}_y$ in a planar layer $-L < z < L$ in infinite fluid one finds a flow in the x direction given by

$$\begin{aligned} v_x(z) &= U, \qquad L < z < \infty, \\ &= Uz/L, \qquad -L < z < L, \\ &= -U, \qquad -\infty < z < -L, \end{aligned} \tag{4}$$

with $U = LT/2\eta$. This shows that in the absence of a confining boundary a uniform but localized torque density acts like a pump on the fluid. One can visualize the flow as being generated by spinning particles in the layer. A more complete description including the spinning particles leads to a smooth flow profile for non-vanishing spin viscosity[3]. Two parallel layers of opposite and equal torque density generate a uniform plug flow between the two layers. The question is whether a collective motion of this type can be used for practical purposes. As shown in the next two examples, confinement tends to annihilate the flow.

Consider first a fluid confined to a spherical cavity of radius R with no-slip boundary condition at the wall of the cavity. We choose the origin at the center of the cavity and use spherical coordinates (r, θ, φ). A uniform torque density is applied in the spherical region $0 < r < c$, with $c < R$,

$$\boldsymbol{T}(\boldsymbol{r}) = T\boldsymbol{e}_z\Theta(c - r), \tag{5}$$

where $\Theta(x)$ is the Heaviside step-function. The torque density has an effect only at $r = c$, since

$$\nabla\times\boldsymbol{T}(\boldsymbol{r}) = T\boldsymbol{e}_\varphi\delta(r - c). \tag{6}$$

One solves the flow problem by putting

$$\boldsymbol{v}(\boldsymbol{r}) = f(r)\sin\theta\,\boldsymbol{e}_\varphi, \qquad p(\boldsymbol{r}) = p_0, \tag{7}$$

with scalar function $f(r)$. Substituting one finds

$$\begin{aligned} f(r) &= Wr, \qquad 0<r<c, \\ &= Ar + Br^{-2}, \qquad c<r<R, \end{aligned} \tag{8}$$

with coefficients W, A, B. Solving for the coefficients from the jump conditions at $r=c$ and the boundary condition at $r=R$ one finds in particular for the coefficient W

$$W = \frac{R^3-c^3}{3R^3}\frac{T}{\eta}. \tag{9}$$

This shows that the rotational velocity of the inner sphere $r<c$ tends to zero as $c \to R$ due to friction with the wall. For small gap $R-c$ one has approximately

$$W \approx (R-c)\frac{T}{\eta R}, \tag{10}$$

with angular velocity proportional to the width of the gap.

Similarly, for a cylinder of radius R with no-slip condition at the wall of the cylinder one introduces cylindrical coordinates (r,φ,z). Assuming a torque density of the form Eq. (2.5) one solves the flow problem by putting

$$\boldsymbol{v}(\boldsymbol{r}) = f(r)\,\boldsymbol{e}_\varphi, \qquad p(\boldsymbol{r}) = p_0, \tag{11}$$

with scalar function $f(r)$ given by

$$\begin{aligned} f(r) &= Wr, \qquad 0<r<c, \\ &= Ar + Br^{-1}, \qquad c<r<R. \end{aligned} \tag{12}$$

This leads to the coefficient W

$$W = \frac{R^2-c^2}{2R^2}\frac{T}{\eta}. \tag{13}$$

For small gap this is again approximated by Eq. (2.10).

3. Ferrohydrodynamics

In practical applications a torque density can be generated by an electric field acting on a permanent electrical dipole moment, or by a magnetic field acting on a permanent magnetic moment. There must be an angle between field and dipole moment. This can be achieved for example by acting with a rotating field. Then due to dissipation there will be a lag between field and moment, and hence a non-vanishing torque density. We consider the magnetic case. Then one must use the equations of ferrohydrodynamics[1]. The long range interactions between dipole moments are described on the macroscopic level by Maxwell's equations. The lag between magnetization and magnetic field must be found from a magnetic relaxation equation.

In ferrohydrodynamics Stokes' equations are generalized to

$$\nabla\cdot(\boldsymbol{\sigma}_{hyd} + \boldsymbol{\sigma}_m) = 0, \qquad \nabla\cdot\boldsymbol{v} = 0, \qquad \nabla\cdot\boldsymbol{\omega_p} = 0, \tag{14}$$

where $\boldsymbol{\sigma}_{hyd}$ is the hydrodynamic stress tensor, given in cartesian component form by

$$\sigma_{hyd,\alpha\beta} = -p\,\delta_{\alpha\beta} + \eta\left(\frac{\partial v_\alpha}{\partial x_\beta} + \frac{\partial v_\beta}{\partial x_\alpha}\right) + \zeta\,\varepsilon_{\alpha\beta\gamma}(\nabla\times\boldsymbol{v} - 2\boldsymbol{\omega}_p)_\gamma, \tag{15}$$

where $\varepsilon_{\alpha\beta\gamma}$ is the Levi-Civita tensor, $\boldsymbol{\omega}_p$ is the spin velocity of suspended particles, and $\boldsymbol{\sigma}_m$ is the Maxwell stress tensor. The latter is given by

$$\boldsymbol{\sigma}_m = \boldsymbol{BH} - \frac{1}{2}\mu_0 H^2\boldsymbol{I}, \tag{16}$$

in SI units, where $\boldsymbol{B}$ is the magnetic induction, μ_0 is the magnetic permeability of vacuum, $\boldsymbol{H}$ is the magnetic field, and $\boldsymbol{I}$ is the unit tensor. Induction and field are related by

$$\boldsymbol{B} = \mu_0(\boldsymbol{H} + \boldsymbol{M}), \tag{17}$$

where $\boldsymbol{M}$ is the magnetization, and are assumed to satisfy Maxwell's equations of magnetostatics,

$$\nabla\cdot\boldsymbol{B} = 0, \qquad \nabla\times\boldsymbol{H} = \boldsymbol{j}, \tag{18}$$

where $\boldsymbol{j}$ is the externally applied current density. The divergence of the Maxwell stress tensor yields both a magnetic force density and a magnetic torque density, acting on the fluid. The first equation (3.1) is the equation of motion for momentum, with neglect of inertial terms. We must also consider the equation of motion for the spin velocity $\boldsymbol{\omega}_p$, or equivalently for the spin per unit mass $\boldsymbol{S} = I\boldsymbol{\omega}_p$, where I is a moment of inertia per unit mass, with dimension a^2, where a is a radius of gyration. The equation of motion for spin reads

$$\rho\frac{d\boldsymbol{S}}{dt} = 2\zeta(\nabla\times\boldsymbol{v} - 2\boldsymbol{\omega}_p) + \mu_0\boldsymbol{M}\times\boldsymbol{H} + \eta'\nabla^2\boldsymbol{\omega}_p, \tag{19}$$

where ρ is the mass density and η' is the spin viscosity, which may be interpreted as a spin diffusion coefficient. The second term on the right is the magnetic torque density. On a slow time scale we may drop the inertial terms on the left, like in the momentum equation, and are then left with

$$\eta'\nabla^2\boldsymbol{\omega}_p + 2\zeta(\nabla\times\boldsymbol{v} - 2\boldsymbol{\omega}_p) + \mu_0\boldsymbol{M}\times\boldsymbol{H} = 0. \tag{20}$$

Substituting this into Eq. (3.2) we find that the first equation (3.1) may be rewritten as

$$\eta\nabla^2\boldsymbol{v} - \nabla p = -\boldsymbol{F}^s_m - \frac{1}{2}\eta'\,\nabla\times(\nabla^2\boldsymbol{\omega_p}), \tag{21}$$

where $\boldsymbol{F}^s_m$ is the magnetic force density, which follows from the symmetric part of the Maxwell stress tensor as

$$\boldsymbol{F}^s_m = \nabla\cdot\boldsymbol{\sigma}^s_m = \mu_0\boldsymbol{M}\cdot(\nabla\boldsymbol{H}) + \frac{1}{2}\mu_0\nabla\times(\boldsymbol{M}\times\boldsymbol{H}). \tag{22}$$

The first term on the right is the Kelvin force density. The second term may be combined with the last term in Eq. (3.8) to yield a torque density $\mu_0\boldsymbol{M}\times\boldsymbol{H} + \eta'\nabla^2\boldsymbol{\omega}_p$, as in Eq. (2.2). Thus the equation of motion (3.8) for the flow velocity $\boldsymbol{v}$ is quite similar to the equation we have studied in the preceding section.

The equations of motion (3.7) and (3.8), together with the condition of incompressibility $\nabla\cdot\boldsymbol{v} = 0$, must be supplemented with a constitutive equation for the magnetization. In practice one uses a relaxation equation with a single relaxation time of the form

$$\frac{\partial\boldsymbol{M}}{\partial t} + \boldsymbol{v}\cdot\nabla\boldsymbol{M} - \boldsymbol{\omega}_p\times\boldsymbol{M} = -\gamma(\boldsymbol{M} - \boldsymbol{M}_{eq}(\boldsymbol{H})), \tag{23}$$

where γ is a relaxation rate, and $\boldsymbol{M}_{eq}(\boldsymbol{H})$ is the equilibrium magnetization corresponding to the local magnetic field according to the equilibrium equation of state at the local temperature. Heat conduction is assumed to be fast, so that the temperature can be taken to be uniform. The relaxation rate γ can be identified with $1/\tau$, where τ is the relaxation time for rotation of a particle. The suspension is usually polydisperse, and the longest relaxation time, corresponding to the largest particles will dominate. The actual relaxation process will be more complicated, but Eq. (3.10) captures the main features, and is suitable for exploratory calculations. The equation introduces a time-dependence into the problem. Moreover it is nonlinear, so that a perturbation method must be used.

4. Collective motion of a ferrofluid

In a ferrofluid one can generate collective motion by the action of a time-dependent applied field. It is convenient to take the field to be periodic in time, and to ask for a systematic collective motion of the fluid on time average. The early experiments and theory have been concerned with rotation of a ferrofluid in a cylindrical vessel under the influence of a rotating magnetic field. Later, pumping by a running wave magnetic field in planar or cylindrical geometry has been considered.

Rotation of a ferrofluid in a cylindrical vessel by means of a rotating magnetic field was achieved first by Rosensweig and Moskowitz[6]. A theoretical explanation of bulk rotational motion was provided by Zaitsev and Shliomis[7]. They found that the ferrofluid rotates with the field at a rate proportional to the magnetic field. Experimentally the situation is confused. Brown and Horsnell[8] repeated the experiment, and found rotation in the opposite direction. It was realized later that in experiments with an open cylinder, the coupling to a surface wave dominates, and that one can get both directions of rotation, depending on whether the surface is convex or concave[9],[10].

More recently, experiments with a closed cylinder have been performed[11],[12], in order to eliminate the surface wave effect. On the basis of these experiments Chaves et al. concluded to a value of the spin viscosity of about $\eta' \approx 10^{-12} - 10^{-8}$ Ns. This is much larger than Rosensweig's earlier estimate $\eta' \approx \eta d^2$, where d is the mean distance between suspended particles. The estimate leads to a value $\eta' \approx 10^{-18}$ Ns at volume fraction $\phi = 0.1$, much too small to be compatible with observed rotation rates.

As a test, the rotation experiment has been repeated with a closed sphere filled with ferrofluid[13]. Surprisingly, the fluid did not rotate noticeably in the rotating field. A theoretical estimate with the value of the spin viscosity found by Chaves et al. leads to a rotation rate which should have been observable. Khushrushahi and Zahn[13] conclude that the earlier analysis[12] led to an incorrect value due to neglect of a non-uniform demagnetization field. In the analogous situation of an electrically polar liquid Bonthuis et al.[14] estimated the spin viscosity as $\eta' \approx \eta a^2$, where a is a molecular radius. This agrees with computer simulation results for nonpolar molecular liquids of Evans and Streett[15]. Hansen et al.[16] found, on the basis of molecular dynamics simulations and a theoretical expression of the spin viscosity as the integral of a time-correlation function, a value for η' two orders of magnitude larger than this estimate. This suggests that the spin viscosity is affected significantly by long range dipolar interactions.

Further theoretical analysis of the rotation experiment has been based on the equations of the preceding section. As a starting point one performs perturbation theory in the amplitude of the applied field. It is assumed that the applied field and the resulting magnetization are much smaller than the saturation magnetization of the ferrofluid, so that the equilibrium of state in Eq. (3.10) may be linearized as $\boldsymbol{M}_{eq}(\boldsymbol{H}) \approx \chi_0 \boldsymbol{H}$, where χ_0 is the initial susceptibility. We consider a uniform rotating applied field with x, y components

$$H_{1x}(t) = A\cos\omega t, \qquad H_{1y} = A\sin\omega t, \tag{24}$$

where the subscript 1 indicates that the field is linear in the amplitude A. A perturbation analysis in powers of the amplitude A then takes the form

$$\begin{aligned} \boldsymbol{H} &= \boldsymbol{H}_1 + \boldsymbol{H}_3 + ..., & \boldsymbol{M} = \boldsymbol{M}_1 + \boldsymbol{M}_3 + ..., \\ \boldsymbol{v} &= \boldsymbol{v}_2 + \boldsymbol{v}_4 + ..., & p = p_0 + p_2 + p_4 + ..., \\ \boldsymbol{\omega}_p &= \boldsymbol{\omega}_{p2} + \boldsymbol{\omega}_{p4} + ..., \end{aligned} \tag{25}$$

where the subscript denotes the power of A. To lowest order one evaluates ($\boldsymbol{M}_1$, $\boldsymbol{H}_1$, $\boldsymbol{v}_2$, $\boldsymbol{\omega}_{p2}$). The field $\boldsymbol{H}_1 = (H_{1x}, H_{1y}, H_{1z})$ is given by Eq. (4.1). The magnetization $\boldsymbol{M}_1$ is harmonic as well, but lags the field,

$$M_{1x}(t) = A\cos(\omega t - \alpha), \qquad M_{1y} = A\sin(\omega t - \alpha). \tag{26}$$

The lag angle α is expressed in terms of the complex susceptibility $\chi(\omega) = \chi'(\omega) + i\chi''(\omega)$ as

$$\alpha = \arctan\frac{\chi''(\omega)}{\chi'(\omega)}. \tag{27}$$

The susceptibility follows from Eq. (3.10) and $\boldsymbol{M}_{1\omega} = \chi(\omega)\boldsymbol{H}_{1\omega}$ as

$$\chi(\omega) = \frac{\gamma\chi_0}{\gamma - i\omega}. \tag{28}$$

We use complex notation with time-factor $\exp(-i\omega t)$. Hence the lag angle is

$$\alpha = \arctan\frac{\omega}{\gamma}. \tag{29}$$

In the experiment one has $\omega << \gamma$, so that $\alpha \approx \omega/\gamma$. To second order the magnetic torque density is

$$\boldsymbol{T}_2 = \mu_0\boldsymbol{M}_1 \times \boldsymbol{H}_1. \tag{30}$$

This is uniform and independent of time

$$\boldsymbol{T}_2 = T_{2z}\boldsymbol{e}_z, \qquad T_{2z} = \mu_0\chi''(\omega)A^2, \tag{31}$$

with

$$\chi''(\omega) = \chi_0\frac{\omega\gamma}{\omega^2 + \gamma^2} \approx \chi_0\frac{\omega}{\gamma}. \tag{32}$$

The torque density vanishes at zero frequency, since then the lag angle α vanishes.

To second order Eqs. (3.7) and (3.8) become

$$\begin{aligned} \eta\nabla^2\boldsymbol{v}_2 - \nabla p_2 &= -\frac{1}{2}\eta'\,\nabla \times (\nabla^2\boldsymbol{\omega}_{p2}), \\ \eta'\nabla^2\boldsymbol{\omega}_{p2} + 2\zeta(\nabla \times \boldsymbol{v}_2 - 2\boldsymbol{\omega}_{p2}) &= -\boldsymbol{T}_2. \end{aligned} \tag{33}$$

The force density $\boldsymbol{F}^s_{m2}$ vanishes to this order, since $\boldsymbol{M}_1$ and $\boldsymbol{H}_1$ are uniform. The equations can be solved for spherical[17], and cylindrical geometry[18], and lead to steady-state flow velocity $\boldsymbol{v}_2$ and spin velocity $\boldsymbol{\omega}_{p2}$. For a cylinder the solution is similar to that of Zaitsev and Shliomis[7], but differs because these authors used the unsatisfactory approximation $\boldsymbol{M}_{eq}(\boldsymbol{H}) \approx M_s\boldsymbol{H}/H$, where M_s is the saturation magnetization.

For a sphere the flow velocity $\boldsymbol{v}_2(\boldsymbol{r})$ takes the form (2.7), and for a cylinder it takes the form (2.11). Correspondingly, for a sphere the spin velocity takes the form

$$\boldsymbol{\omega}_{p2}(\boldsymbol{r}) = \nabla \times (g(r)\sin\theta\,\boldsymbol{e}_\varphi), \tag{34}$$

with radial function $g(r)$, whereas for a cylinder one has simply

$$\boldsymbol{\omega}_{p2}(\boldsymbol{r}) = g(r)\boldsymbol{e}_z. \tag{35}$$

As an example, for a sphere the solution takes the form

$$f(r) = Wr + XF(\kappa r), \qquad g(r) = \frac{T_{2z}}{8\zeta} + \frac{1}{2}Wr + \frac{\eta+\zeta}{2\zeta}XF(\kappa r), \tag{36}$$

with the function

$$F(x) = \frac{\cosh x}{x} - \frac{\sinh x}{x^2}, \tag{37}$$

and inverse screening length

$$\kappa = \sqrt{\frac{4\eta\zeta}{\eta'(\eta+\zeta)}}. \tag{38}$$

The coefficients W and X are determined from the boundary conditions $\boldsymbol{v} = 0$ and $\omega_{p\varphi} = 0$ at $r = R$.

For a cylinder the solution of Eqs. (3.7), (3.8), and (3.10) can be carried out to all orders in the amplitude A in a numerical iterative scheme in a first harmonic approximation, keeping only steady-state terms and first harmonics, and with linearized equilibrium equation of state[18]. The second order perturbation theory result is used as a starting point in the iteration. Surprisingly, one finds that for a large amplitude the solution bifurcates into two solutions with rotation in opposite directions. Experimentally this corresponds to flow reversal at strong applied field, since the solution with rotation opposite to the field has the lower dissipation. Such flow reversals have been observed experimentally, but in some experiments the surface wave phenomenon may have dominated.

5. Discussion

For simplicity we have restricted attention to a rotating applied magnetic field of the form (4.1). This is the simplest example of collective motion generated by a varying applied magnetic field. In order to achieve pumping of a ferrofluid more complicated variations in space and time will be required. Solutions which have been considered involve plane magnetic waves running along a planar duct or a circular tube[19]-[21]. It will be important to study other situations, and find improved ways of technical interest.

The spin diffusion described by the transport coefficient spin viscosity leads to a smooth boundary layer. Theoretical estimates of the spin viscosity and the experiment of Khushrushahi and Zahn[13] suggest that the boundary layer is quite thin, but more work on the value of the transport coefficient is desirable.

The boundary condition used for the spin velocity is not well established. It is known that a single particle, confined in a spherical cavity filled with viscous incompressible fluid and subjected to a constant torque, shows circular orbital motion, with reversal of direction for orbits close to the surface[22]. It would be of interest to find the motion of a small number of particles confined to a spherical cavity or a circular tube by Stokesian dynamics simulation. The transition to collective motion and the behavior near the boundary may provide information on the role of spin viscosity and the proper boundary condition to be used in ferrohydrodynamics.

References

[1] Rosensweig R E 1985 *Ferrohydrodynamics* (Cambridge University Press, Cambridge)
[2] Mao L, Elborai S, He X, Zahn M, and Koser H 2011 *Phys. Rev. B* vol 84, 104431
[3] Felderhof B U 2012 *J. Chem. Phys.* vol 136, 164905

[4] Happel J and Brenner H 1973 *Low Reynolds number hydrodynamics* (Noordhoff, Leyden)
[5] Felderhof B U 1976 *Physica A* vol 82, 596
[6] Moskowitz R and Rosensweig R E 1967 *Appl. Phys. Lett.* vol 11, 301
[7] Zaitsev V M and Shliomis M I 1969 *J. Appl. Mech. Tech. Phys.* vol 10, 696
[8] Brown R and Horsnell T S 1969 *Electr. Rev.* 235
[9] Mailfert R and Martinet A 1973 *J. de Physique* vol 34, 197
[10] Rosensweig R E, Popplewell J, and Johnston R J 1990 *J. Magn. Magn. Mat.* vol 85, 171
[11] Chaves A, Rinaldi C, Elborai S, He X, and Zahn M 2006 *Phys. Rev. Lett.* vol 96, 194501
[12] Chaves A, Zahn M, and Rinaldi C 2008 *Phys. Fluids* vol 20, 053102
[13] Khushrushahi S and Zahn M 2011 *J. Magn. Magn. Mat.* vol 323, 1302
[14] Bonthuis D J, Horinek D, Bocquet L, and Netz R 2009 *Phys. Rev. Lett.* vol 103, 144503
[15] Evans D J and Streett W B 1978 *Mol. Phys.* vol 36, 161
[16] Hansen J S, Bruus H, Todd B D, and Daivis P J 2010 *J. Chem. Phys.* vol 133, 144906
[17] Felderhof B U 2011 *Phys. Rev. E* vol 84, 046313
[18] Felderhof B U 2011 *Phys. Rev. E* vol 84, 026312
[19] Felderhof B U 2011 *Phys. Fluids* vol 23, 042001
[20] Felderhof B U 2011 *Phys. Fluids* vol 23, 092003
[21] Felderhof B U 2011 *Phys. Fluids* vol 23, 092002; erratum 2012 *Phys. Fluids* vol 24, 019901
[22] Jones R B 2009 in *Theoretical Methods for Microscale Viscous Flows*, eds. Feuillebois F and Sellier A (Transworld Research Network, Kerala, India), p 61

Sedimentation of non-Brownian suspensions

G C Abade[1] **and B Cichocki**[2]

[1] Departamento de Engenharia Mecânica, Universidade de Brasília, Campus Darcy Ribeiro, 70910-900, Asa Norte, Brasília-DF, Brazil

[2] Institute of Theoretical Physics, University of Warsaw, Hoża 69, 00-681 Warsaw, Poland

E-mail: abade@unb.br

Abstract. We describe the sedimentation of a non-Brownian suspension of hard spheres by Stokesian dynamics simulations. Attention is focused on the mean settling velocity and the structure of particles distribution as the suspension attains a non-equilibrium steady sedimentation state. Many-body hydrodynamic interactions between particles are evaluated using the precise multipole method, encoded in the new FAST-HYDROMULTIPOLE program. With the complete neglect of Brownian motion, the sedimentation velocity and the particles configurational distribution in the steady state differ measurably from their equilibrium counterparts. Simulations reproduce short-range particle correlations in good agreement with the statistical description of sedimentation formulated by Cichocki and Sadlej [Europhys. Lett. **72**, 936 (2005)].

1. Introduction

We address the sedimentation of a dilute suspension of non-Brownian spheres, with special attention being paid to the mean sedimentation velocity and the underlying spatial distribution of the spheres throughout the fluid.

When the randomizing Brownian motion is absent, the dynamics of sedimenting spheres is governed by the interplay of permanent particle transport (driven by external forces) and the complex long-range hydrodynamic interactions mediated by the background fluid. This purely Stokesian dynamics poses severe difficulties to the determination of the suspension microstructure and the corresponding observable suspension properties, such as the average sedimentation velocity.

The average settling velocity is characterized by the sedimentation coefficient U/U_0, where U is the mean sedimentation speed and U_0 is the settling velocity of an isolated sphere. For dilute suspensions, the sedimentation coefficient is represented in terms of an expansion in powers of the particle volume fraction ϕ,

$$\frac{U}{U_0} = 1 + \lambda\phi + O(\phi^2). \tag{1}$$

Calculation of the linear coefficient λ involves the evaluation of hydrodynamic interactions and presupposes a pair correlation function $g(\mathbf{r})$ describing the suspension microstructure. This pair correlation function may be assumed or derived.

Batchelor [1] obtained $\lambda = -6.55$ by assuming a pair correlation function that corresponds to the equilibrium particle distribution, where all configurations of non-overlapping spheres are equally probable. Batchelor's assumption would be plausible if the Brownian motion played a dominant role in determining the particles distribution.

The non-Brownian case was tackled by Cichocki and Sadlej [2, 3, 4]. Instead of assuming *a priori* an equilibrium structure, they predicted the non-equilibrium pair correlation function using statistical-mechanical considerations, providing $\lambda = -3.87$ [2]. This result is in excellent agreement with the value -3.9, obtained by fitting Eq. (1) to the experimental data [5, 6] for the settling velocity of non-Brownian suspensions in the dilute regime. For example, at a volume fraction $\phi = 0.05$, Batchelor's result predicts $U/U_0 \approx 0.67$, while Cichocki and Sadlej provide $U/U_0 \approx 0.81$. The correction to the average settling velocity arising from relaxation of the particle distribution away from equilibrium is quite considerable.

In this paper we describe, by computer simulations of the Stokesian dynamics, the relaxation of the mean sedimentation velocity as the system develops a non-equilibrium microstructure. To describe the dynamics in the bulk of a suspension which is spatially uniform on the macroscopic scale, we follow an established practice and impose periodic boundary conditions. Hydrodynamic interactions between many spheres are properly evaluated using the FAST-HYDROMULTIPOLE code [7], an accelerated version of the well-established HYDROMULTIPOLE program [8, 9, 10, 11]. The method underlying both numerical codes is based on a multipole moment expansion of the formal integral solution of the Stokes equations. In the FAST-HYDROMULTIPOLE, the speed of calculation is improved by use of the Fast Multipole Method (FMM), developed by Greengard and Rokhlin [12] and widely used for efficient computation of gravitational and electrostatic interactions. For hydrodynamic interactions, the FMM allows for evaluation of the velocity fields generated by a collection N particles in $O(N)$ operations.

This article is organized as follows. Section 2 describes our model system and the equations governing the Stokesian dynamics of suspensions. The numerical scheme for calculation of hydrodynamic interactions is outlined. Section 3 presents simulation results for the mean sedimentation velocity and positional particle correlations in a dilute suspension. Section 4 contains discussions.

2. Model system and simulation method

We consider a collection of N hard spheres, each of radius a, immersed in an unbounded viscous incompressible fluid of shear viscosity η. The spheres are made to move through the fluid by an externally applied constant force $\boldsymbol{F}$ acting on each of the particles. The motion of the immersed spheres and the suspending fluid are coupled by no-slip velocity boundary condition at the fluid-particle interfaces. The set up fluid flow is assumed to be described by the stationary Stokes equations,

$$\eta\nabla^2\boldsymbol{v} - \nabla p = 0, \qquad \nabla\cdot\boldsymbol{v} = 0, \tag{2}$$

where $\boldsymbol{v}$ is the fluid velocity and p is the pressure field.

The suspended spheres, centered at $\mathbf{X} = (\boldsymbol{R}_1, \dots, \boldsymbol{R}_N)$, undergo a rigid body motion with translational velocities $\mathbf{U} = (\boldsymbol{U}_1, \dots, \boldsymbol{U}_N)$. It follows from linearity of Eq. (2) and boundary conditions that particle velocities $\mathbf{U}$ are linearly related to the driving forces $\mathbf{F} = (\boldsymbol{F}, \dots, \boldsymbol{F})$,

$$\frac{\mathrm{d}\mathbf{X}}{\mathrm{d}t} = \mathbf{U} = \boldsymbol{\mu}(\mathbf{X})\mathbf{F}, \tag{3}$$

where $\boldsymbol{\mu}$ is the $3N \times 3N$ translational mobility matrix depending on the instantaneous configuration $\mathbf{X}$. The mobility matrix formally embodies all information on the fluid mediated hydrodynamic interactions. Next we outline the scheme for proper account of these interactions.

2.1. The mobility matrix

The N-particle mobility matrix $\boldsymbol{\mu}$ may be expressed formally as follows [13]:

$$\boldsymbol{\mu} = \boldsymbol{\mu}_0 + \boldsymbol{\mu}_0 \mathbf{Z}_0 \sum_{k=0}^{\infty} (-\mathbf{G}\hat{\mathbf{Z}}_0)^k \mathbf{G}\mathbf{Z}_0 \boldsymbol{\mu}_0, \tag{4}$$

where $\boldsymbol{\mu}_0 = (6\pi\eta a)^{-1}\mathbf{1}$ is the mobility of an isolated sphere and $\mathbf{1}$ is the unit matrix. The second term on the right-hand side of Eq. (4) accounts for the hydrodynamic interactions. The hydrodynamic operators $\mathbf{Z}_0$, $\hat{\mathbf{Z}}_0$, and $\mathbf{G}$ are N-particle matrices with

$$[\mathbf{Z}_0]_{ij} = \mathbf{Z}_0(i)\delta_{ij}, \qquad [\mathbf{G}]_{ij} = \mathbf{G}(ij)(1-\delta_{ij}), \qquad i,j = 1,\ldots,N, \tag{5}$$

where $\mathbf{Z}_0(i)$ and $\mathbf{G}(ij)$ are, respectively, single- and two-particle integral operators. $\hat{\mathbf{Z}}_0$ has the same single-particle structure as $\mathbf{Z}_0$. The hydrodynamic operators will be defined below (details may be found e.g. in Ref. [11]).

When acted on by an external force $\boldsymbol{F}$, a source particle j exerts on the fluid a force density, distributed along the fluid-particle interface. This force density gives rise to a disturbing flow field, which is propagated and eventually collides with the other suspended particles. The two-particle Green integral propagator $\mathbf{G}(ij)$ appearing in Eq. (5) describes the incident velocity field on the target particle i due to a force density applied on the fluid by the source particle j.

When immersed in the incident fluid flow generated by the suspended particles and acted on by $\boldsymbol{F}$, a test particle i resists the flow; it exerts on the fluid a force density, which produces a reflecting flow field. The object $\mathbf{Z}_0(i)$ (Eq. (5)) gives the force density on the surface of particle i in response to a given incident field. Finally the operator $\hat{\mathbf{Z}}_0$ is the *convective* counterpart of $\mathbf{Z}_0$, i.e. corresponds to the situation when the immersed particle is convected by the incident flow without exerting net force or torque on the fluid.

In Equation (4), the term

$$\sum_{k=0}^{\infty} (-\mathbf{G}\hat{\mathbf{Z}}_0)^k = \mathbf{1} - \mathbf{G}\hat{\mathbf{Z}}_0 + \mathbf{G}\hat{\mathbf{Z}}_0\mathbf{G}\hat{\mathbf{Z}}_0 - \cdots \tag{6}$$

describes the infinite sequence of reflected (or scattered) flow fields. The scattering series (6) is responsible for the full account of the many-body hydrodynamic interactions.

2.2. Fast evaluation of particle velocities

Fast schemes for Stokesian dynamics simulations avoid explicit calculation of the $3N \times 3N$ mobility matrix $\boldsymbol{\mu}$. Evaluation of the $3N$-dimensional vector of particle velocities $\mathbf{U}$ in response to known driving forces $\mathbf{F}$ suffices to generate particle trajectories according to (3).

Recalling Eq. (3), the determination of particle velocities $\mathbf{U}$ is done by applying the mobility matrix to $\mathbf{F}$. This operation entails several matrix-vector multiplications involving $\mathbf{Z}_0$, $\hat{\mathbf{Z}}_0$ and $\mathbf{G}$ (see (4) and (6)). The single-particle matrices, $\mathbf{Z}_0$ and $\hat{\mathbf{Z}}_0$, are diagonal in particle indices, and their application to an $3N$-dimensional vector is a fast $O(N)$ operation. The N-particle matrix $\mathbf{G}$ of propagators (defined in Eq. (5)) has however $O(N^2)$ nonzero elements.

An accelerated scheme for evaluating particle velocities is based on a sparse representation of the matrix $\mathbf{G}$ suitable for fast computation of matrix-vector multiplication. This representation is constructed by employing the ideas underlying the Fast Multipole Method (FMM) [12].

Application of the FMM to hydrodynamic interactions [7] yields a formal representation of matrix $\mathbf{G}$ in the form

$$\mathbf{G} = [\mathbf{G}]_N + \mathbf{S}^{\dagger}\tilde{\mathbf{G}}\mathbf{S}, \tag{7}$$

where $[\mathbf{G}]_N$ is a restriction of $\mathbf{G}$ which includes only two-particle propagators $\mathbf{G}(ij)$ between nearest-neighbours, and thereby contains $O(N)$ nonzero elements. The term $\mathbf{S}^\dagger\tilde{\mathbf{G}}\mathbf{S}$ is formally an $O(N)$ sparse matrix accounting for hydrodynamic interactions between the remaining well-separated particles. Matrix $\tilde{\mathbf{G}}$ is formed by Green propagators connecting well-separated groups of particles, arranged at different length scales in a self-similar ($\mathbf{X}$-independent) structure. Hence $\tilde{\mathbf{G}}$ is completely decoupled from the details of the configuration of particles centres and may be pre-calculated. The operator $\mathbf{S}$ and its adjoint $\mathbf{S}^\dagger$ (defined in Ref. [7]) are responsible for this decoupling. Details of the numerical implementation of this scheme will be published separately.

2.3. Multipole method

Numerical calculation of the mobility matrix requires transformation of the integral operators $\mathbf{Z}_0$, $\hat{\mathbf{Z}}_0$ and $\mathbf{G}$ (Eq. (4)) into algebraic ones. Transformation is done by projecting and expanding them in spherical multipole functions around particle centers (see e.g. Refs. [10, 11, 14] for details). Multipole functions are indexed by $l = 1, 2, \ldots$, $m = -l, \ldots, l$, and $\sigma = 0, 1, 2$. Explicit expressions for the multipole elements of the hydrodynamic operators may be found in Ref. [14].

It follows from the multipole expansion of the Green propagator [15] that

$$(lm\sigma|\mathbf{G}(\mathbf{R})|l'm'\sigma') \sim \frac{1}{R^{l+l'+\sigma+\sigma'-1}}, \tag{8}$$

where $\mathbf{R}$ is the relative distance vector of two centers of expansion. These may be the centers of two interacting spheres or two interacting groups of particles in the fast scheme.

This method has the distinguishing characteristic that the multipole functions are expressed in terms of an irreducible representation. Therefore Eq. (8) furnishes a complete insight into the structure of the spatial decay of interactions between point multipoles of different orders (labeled by $lm\sigma$ and $l'm'\sigma'$). To account for essential collective effects due to hydrodynamic interactions, a reasonable prescription is to keep all long-range multipole matrix elements of the propagator $\mathbf{G}(\mathbf{R})$, i.e. decaying as slowly as $1/R^3$ [10, 16]. This is done by including all multipoles with $l \leq 3$.

Summarizing, Eqs. (3), (4), and (7), with an irreducible multipole representation of the hydrodynamic operators, form the basis of the FAST-HYDROMULTIPOLE simulation tool. Our treatment combines proper high-order multipole truncation and a full account of many-body hydrodynamic interactions through the multiple scattering series (6). This account includes all essential effects to assure the nonvanishing of the divergence of particle velocities in the Stokes-Liouville equation (see e.g. Ref. [2]), the statistical-mechanical equivalent of (3). Important lubrication effects are also included as explained in Ref. [17].

2.4. Simulations

The Stokesian dynamics equations (3) were numerically integrated using the fourth-order Adams-Bashforth algorithm [18]. Since inherent numerical errors in the solution of Eq. (3) by finite-difference integration algorithms may yield unrealistic overlaps between sedimenting spheres, we have devised an alternative scheme for stepping in time particle configurations containing close spheres in relative motion. In our approach, the numerical integration algorithm is supplemented with the physical information that hydrodynamic lubrication forces (which diverge at contact) prevent close spheres from touching each other. The proposed method is fast and effectively prevents particle overlaps without the requirement of prohibitively small integration time-steps. We will provide a detailed description of this algorithm in a separate publication.

Simulations were carried out for the volume fraction $\phi = 0.05$ and for two system sizes, $N = 512$ and 1024. Integration of Eq. (3) considered a time-step $\Delta t = 0.05a/U_0$ (where U_0 is the settling velocity of an isolated sphere). The time evolution started from two different initial distributions of non-overlapping spheres: (i) equilibrium with periodic boundary conditions and (ii) random distribution of hard spheres inside a box with impenetrable walls. The condition of impenetrable walls is imposed only for the purpose of generating an initial configurational distribution of the particles which is different from the equilibrium one. Stokesian dynamics simulations in this work were carried out under periodic boundary conditions only. In this way one may verify if the system evolves to the same stationary state, independent of initial conditions. Time evolutions were carried out until measurements of averaged variables were deemed to have reached steady values.

3. Results

3.1. Evolution of the mean settling velocity

This section presents results for the time evolution of the mean sedimentation velocity as the sedimenting suspension develops a non-equilibrium structure.

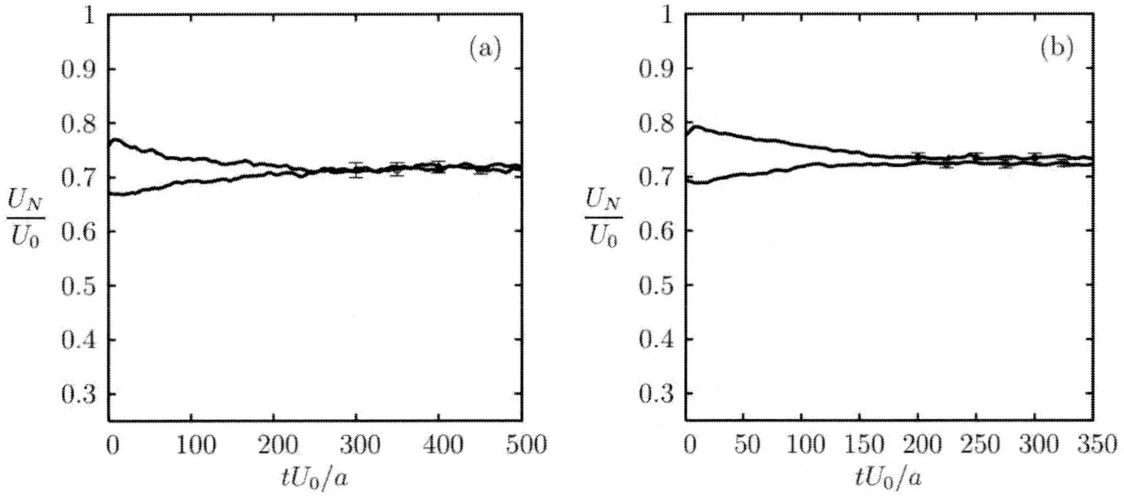

Figure 1. Evolution of the N-particle average sedimentation velocity for the volume fraction $\phi = 0.05$ and two system sizes: (a) $N = 512$; (b) $N = 1024$. The lower curves correspond to simulations starting from equilibrium distributions.

Figure 1 shows the evolution of the N-particle mean sedimentation velocity for $N = 512$ [Fig. 1(a)] and $N = 1024$ [Fig. 1(b)], starting from two different initial particle distributions. The lower curves correspond to simulations starting from equilibrium configurations. For each system size studied a stationary sedimentation state has been established, being this state independent (within statistical errors) of the initial conditions considered. For $N = 512$, the settling velocity attains a steady value after an evolution time of 300 a/U_0 approximately. For the system comprised of 1024 particles, the sedimentation velocity has been relaxed for times larger than $200a/U_0$.

We emphasize that the sedimentation velocity U_N/U_0 plotted in Fig. 1 is for a finite N-particle system. It exhibits considerable system-size dependence due to periodicity (values of U_N/U_0 for $N = 1024$ are slightly larger than those for $N = 512$). To compare with experiments and theory, U_N/U_0 must be extrapolated to its infinite-system counterpart, U/U_0. Extrapolation is performed according to [19]

$$\frac{U}{U_0} = \frac{U_N}{U_0} + 1.76S(0)\frac{\eta}{\eta_{\text{eff}}}\left(\frac{\phi}{N}\right)^{1/3}. \tag{9}$$

Application of Eq. (9) requires as inputs the static structure factor at zero wavenumber, $S(0)$, and the effective suspension viscosity, $\eta_{\rm eff}$. Both $S(0)$ and $\eta_{\rm eff}$ depend on the suspension microstructure. As an approximation, we use Einstein's expression for the suspension viscosity [20], $\eta_{\rm eff}/\eta = 1 + 2.5\phi + O(\phi^2)$. For the steady-state structure factor, we employ the result $S(0) = 1 - 1.64\phi + O(\phi^2)$ derived by Cichocki and Sadlej [2] from their short-range pair correlation function in the steady state. For equilibrium microstructures, $S(0)$ may be evaluated by the Carnahan-Starling expression [19].

Finally we remark that the mean settling rate increases as the suspension evolves away from the equilibrium, i.e. as the suspension structure relaxes towards the non-equilibrium stationary particle distribution. After extrapolation according to Eq. (9), simulations provide the value $U/U_0 = 0.78 \pm 0.01$ for the mean sedimentation velocity in the stationary state, which is larger than sedimentation coefficient $U/U_0 = 0.72 \pm 0.01$ for an equilibrium distribution at $\phi = 0.05$. The uncertainty (± 0.01) in the stationary value of the mean settling velocity was estimated by use of the "blocking" method [21].

3.2. Particle correlations in the steady-state

In this section we present simulation results for the steady-state pair correlation function, $g(\mathbf{r})$.

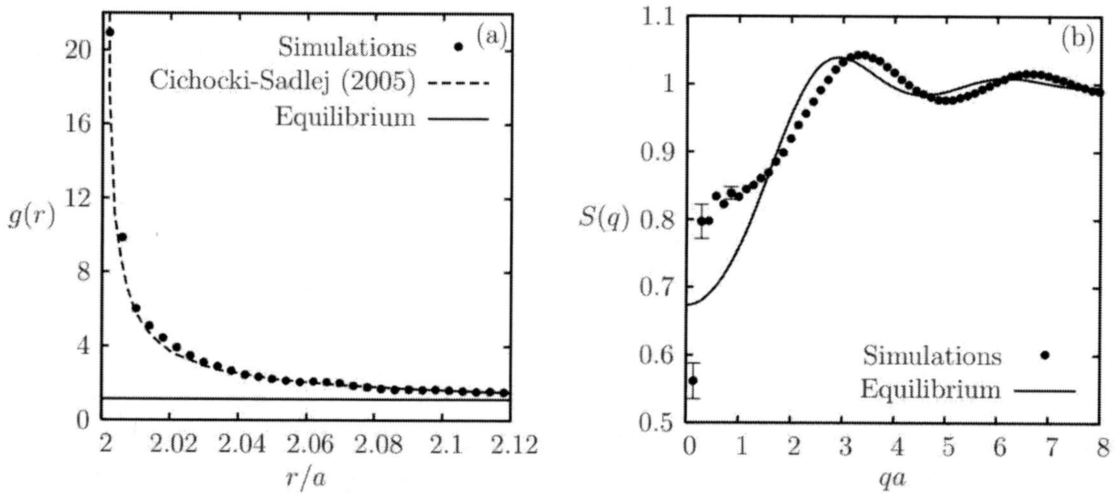

Figure 2. Particle correlations for a steadily sedimenting suspension. (a) The simulated pair correlation function in the steady sedimentation state (solid symbols) compared with its equilibrium counterpart (solid line) at the same volume fraction $\phi = 0.05$. The dashed line represents the stationary-state theoretical result by Cichocki and Sadlej [2]. Both steady-state and equilibrium pair correlation functions vanish for $r/a < 2$. (b) Structure factor for a steadily sedimenting suspension (solid symbols) being compared with the equilibrium distribution (solid line) of hard spheres.

Figure 2(a) shows the spherically averaged numerical results for $g(r)$ in the stationary state in comparison with its equilibrium counterpart, $g_{\rm eq}(r)$. It may be observed that sedimentation produces significant changes in pair correlations relative to equilibrium. It induces the formation of close pairs, which tend to stay together for long times. Prevalence of close pairs explains the increase in the mean settling rate (shown in Fig. 1) as the suspension evolves away from the equilibrium distribution. For our hard-sphere model system, both equilibrium and steady-state pair correlation functions vanish for $r/a < 2$.

Remarkably, Figure 2(a) displays an excellent agreement between our numerical results for $g(r)$ and the isotropic pair correlation function derived theoretically by Cichocki and Sadlej [2] for a dilute suspension (denoted henceforth in the text by $g_{CS}(r)$ for convenience). Agreement with $g_{CS}(r)$ is important since it allows for further exploration of our computer simulation results in the light of the transparent structure of homogeneous sedimentation provided by the Cichocki-Sadlej description. We also point out that our numerical results for $g(r)$ qualitatively agree with the pair distribution obtained by Ladd [22] via large-scale Lattice-Boltzmann simulations. However, Ladd's results were obtained for a larger volume fraction, $\phi = 0.10$.

Further insight into the suspension microstructure at larger length scales is obtained by examining the static structure factor, $S(q)$, related to $g(r)$ by Fourier transform. Figure 2(b) shows our simulation results for the non-equilibrium $S(q)$ in comparison with its equilibrium counterpart, $S_{eq}(q)$. The shape of $S_{eq}(q)$ presented in Fig. 2(b) indicates particle correlations arising from hard-core interactions (recall that $g_{eq}(r) = 0$ for $r < 2a$).

At wavenumbers $qa > 1$, the steady-state data for $S(q)$ obtained from different system sizes and initial particle distributions collapsed into the same curve, with a smooth and well-established wavenumber dependence. At such large wavenumbers, where the short-range structure in a length scale up to $r \sim 6a$ is probed, we were able to obtain an stabilized structure in the stationary state which measurably differs from the equilibrium one. Relative to $S_{eq}(q)$, the position of first peak of $S(q)$ has been slightly displaced to the right, what evidences that sedimentation increases pair correlations in a length scale of the order of the particles diameter.

In the low-q region, however, large fluctuations in $S(q)$ may be found. For a finite system, accurate measurements of the low-q behaviour of $S(q)$ from computer simulation data is particularly troublesome. Considering a simulation of N particles at density n in a periodic cube of side $\mathcal{L} = (N/n)^{1/3}$, any information on the microstructure at long-wavelengths beyond $\mathcal{L}$ will be suppressed by periodicity. Accordingly, $q = 2\pi/\mathcal{L}$ is the smallest accessible wavenumber. Therefore, model systems as large as possible are required to probe the structure in the limit of macroscopic range, i.e. in the limit $q \to 0$ or $\mathcal{L} \to \infty$.

We further discuss and summarize our results in the next section.

4. Discussion

The present work reported on computer simulation results of non-Brownian sedimentation that accompany the Cichocki-Sadlej (C-S) statistical description. The C-S treatment is based on the Stokes-Liouville equation [2] (the statistical-mechanical equivalent of Eq. (3)) in terms of reduced distribution functions describing the dynamics of two hydrodynamically interacting spheres under the intervening influence of a third one. Simulations otherwise follow in detail the full many-particle Stokesian dynamics (3). Consistently, present simulations and the C-S theory share the same careful account of hydrodynamic interactions, using the irreducible multipole description corrected for lubrication effects.

At the studied volume fraction ($\phi = 0.05$), simulations predicted (within statistical uncertainties) a unique non-equilibrium stationary sedimentation state, i.e. independent of the starting particle distribution. With the complete neglect of Brownian motion, the sedimentation velocity and the positional particle correlations in the steady state differ measurably from their equilibrium counterparts.

The short-range pair correlation function in the steady state was found to be in agreement with that predicted by Cichocki and Sadlej [2] (denoted by $g_{CS}(r)$). We have avoided any further analysis on the long-range structure of a sedimenting suspension, since the low-q behaviour of the structure factor $S(q)$ cannot be properly probed from our simulation data. However, the C-S description provides a framework for speculating on the simulated long-range structure by examining the pair correlation function in the short-range. In fact not much is known about the long-range structure of a steadily sedimenting suspension, apart from the fundamental requirement that it must be consistent with the locality of the suspension bulk properties (i.e. they have a well-defined value in the thermodynamic limit). Locality is guaranteed if the pair correlation function $g(\mathbf{r})$ satisfies the Koch-Shaqfeh screening condition [23],

$$n \int d\mathbf{r}[g(\mathbf{r}) - 1] = -1, \tag{10}$$

where $n = N/V$ is the particle number density.

We remark that the condition (10) is a *prescription* underlying the Cichocki-Sadlej derivation of $g_{CS}(r)$. Since $g_{CS}(r)$ fits well to the simulated short-range particle correlations (as Fig. 2 shows), it is tempting to suggest that the simulated microstructure would be consistent with the criterion (10) if the large-scale structures were not cut off at the length of the periodic simulation box. However, this speculation is not conclusive; it must be confirmed by simulations of much larger model systems. For the present, due to system size limitations, agreement with $g_{CS}(r)$ is the only available way to assess consistency of the simulated pair correlation with the Koch-Shaqfeh screening condition (10).

Further analysis of $g_{CS}(r)$ provides an insightful picture on how sedimentation builds up short-range pair correlations in dilute suspensions. As remarked in Ref. [2], the shape of $g_{CS}(r)$ coincides with the pair correlation function calculated by Batchelor and Green [24] for the problem of two spheres immersed in a pure straining flow. This suggests that a pair of close spheres experiences the influence of a third far particle through an effective pure straining flow [2]. Conforming our simulation results to this picture, one may say that simulated long-range structure in the steady state yields an ambient flow pattern reminiscent of a straining fluid motion, which increases pair correlations at small separations and determines the characteristic structure of $g(r)$ represented in Fig. 2(a).

Finally we remark that a fully developed stationary sedimentation state presupposes relaxation of both short and long-range microstructures. Here, under the limitations of a finite-system, attention was restricted to the short-range structure and its correspondent mean sedimentation velocity. Computer simulations to explore the long-range structure and the behaviour of particle velocity fluctuations in the limit of large systems are in progress. We should be able then to investigate if sedimentation produces an anisotropic pair correlation in the long-range. On the theoretical side, a tentative non-equilibrium statistical-mechanical description of particle correlations in the long-range is available [25]. A description of the long-range structure based on the Stokes-Liouville equation still remains to be formulated.

Acknowledgments

This work is part of G.C. Abade's doctoral thesis, elaborated at the University of Warsaw, Institute of Theoretical Physics, under the supervision of Prof. Bogdan Cichocki, and with financial support from CAPES Foundation/Ministry of Education of Brazil. G.C. Abade acknowledges Dr. Maria Ekiel-Jeżewska for her kind invitation and support for participating in the Symposium "Microparticles in Stokes flow" in Honor of François Feuillebois' 65th Birthday. The authors are much indebted to Dr. Eligiusz Wajnryb for indispensable assistance throughout the implementation of the FAST-HYDROMULTIPOLE simulation tool. Valuable discussions with Prof. B. U. Felderhof are gratefully acknowledged. Numerical calculations were carried out at the Centre of Excellence BioExploratorium in Warsaw, Poland.

References

[1] Batchelor G K 1972 *J. Fluid Mech.* **52** 245–268
[2] Cichocki B and Sadlej K 2005 *Europhys. Lett.* **72** 936–942
[3] Cichocki B and Sadlej K 2006 *Arch. Mech.* **58** 327–337
[4] Cichocki B and Sadlej K 2008 *J. Stat. Phys* **132** 129–151

[5] Ham J M and Homsy G M 1988 *Int. J. Multiphase Flow* **14** 533–546
[6] Hanratty T J and Bandukwala A 1957 *AIChE J.* **3** 293–296
[7] Abade G C 2009 *Analysis of transport phenomena in concentrated suspensions by numerical simulations* Ph.D. thesis Warsaw University
[8] Felderhof B U 1988 *Physica A* **151** 1
[9] Cichocki B, Felderhof B U and Schmitz R 1988 *PhysicoChem. Hyd.* **10** 383–403
[10] Cichocki B, Felderhof B, Hinsen K, Wajnryb E and Bławzdziewicz J 1994 *J. Chem. Phys.* **100** 3780
[11] Cichocki B, Jones R B, Kutteh R and E W 2000 *J. Chem. Phys.* **112** 2548–2561
[12] Greengard L and Rokhlin V 1987 *J. Comput. Phys.* **73** 325–348
[13] Szymczak P and Cichocki B 2004 *J. Chem. Phys.* **121** 3329
[14] Wajnryb E, Szymczak P and Cichocki B 2004 *Physica A* **335** 339
[15] Felderhof B U and Jones R B 1989 *J. Math. Phys.* **30** 339
[16] Cichocki B 1996 *Proceedings of the Eighth International Symposium on Continuum Models and Discrete Systems* ed Markov K Z pp 15–20
[17] Cichocki B, Ekiel-Jeżewska M L and Wajnryb E 1999 *J. Chem. Phys.* **111** 3265
[18] Press W H, Teukolsky S A, Vetterling W T and Flannery B P 1992 *Numerical Recipes in Fortran 77, Vol. 1* (Cambridge University Press)
[19] Mo G and Sangani A S 1994 *Phys. Fluids* **6** 1637–1652
[20] Einstein A 1956 *Investigations on the theory of the Brownian motion* (Dover)
[21] Flyvbjerg H and Petersen H G 1989 *J. Chem. Phys.* **91** 461–466
[22] Ladd A J C 1993 *Phys. Fluids A* **5** 299–310
[23] Koch D L and Shaqfeh E S G 1991 *J. Fluid Mech.* **224** 275
[24] Batchelor G K and Green J T 1972 *J. Fluid Mech.* **56** 375–400
[25] Felderhof B U 2005 *Physica A* **348** 16–36

Journal of Physics: Conference Series **392** (2012) 012003 doi:10.1088/1742-6596/392/1/012003

Non-Equilibrium Properties of Semidilute Polymer Solutions under Shear Flow

Chien-Cheng Huang, Gerhard Gompper, and Roland G Winkler

Theoretical Soft Matter and Biophysics, Institute of Complex Systems and Institute for Advanced Simulation, Forschungszentrum Jülich, D-52428 Jülich, Germany

E-mail: `r.winkler@fz-juelich.de`

Abstract. We review results of the non-equilibrium properties of semidilute polymer solutions under shear flow. A hybrid simulation approach is adopted, combing molecular dynamics simulations and the multiparticle collision dynamics method, which has been shown to fully capture hydrodynamic interactions. The polymers exhibit a shear rate dependent deformation and alignment. In addition, shear thinning is observed, which is related to the finite polymer extensibility. It is characterized by the shear-rate dependent viscosity and the first normal-stress coefficient. The conformational and rheological properties are universal functions of a concentration dependent Weissenberg number. The cyclic dynamics of an individual polymer is described in terms of a shear-rate dependent tumbling time. The tumbling time is tightly linked to the polymer end-to-end vector relaxation time and depends on concentration, which is attributed to screening of hydrodynamic interactions in semidilute solutions.

1. Introduction

The properties of dilute polymer solutions under shear flow have been studied intensively [1–18]. Far less attention has been paid to the non-equilibrium properties of semidilute solutions [17–20]. Insight into the behavior of such systems is of fundamental importance in a wide spectrum of systems ranging from biological cells, where transport appears in dense environments, to turbulent drag reduction in fluid flow. While the dynamical behavior of polymers in dilute solution is strongly affected by hydrodynamic interactions [21–23], their relevance in semidilute solutions is less clear.

Experimental studies of single DNA molecules by fluorescence microscopy studies [1–4, 9] reveal a remarkably rich polymer dynamics in shear flow. A polymer chain exhibits large conformational changes and continuously undergoes stretching and compression cycles, denoted as tumbling, and never reaches a steady-state extension. The detailed temporal evolution depends upon the shear rate. These microscopic dynamical properties are tightly linked to the macroscopic rheological behavior of a polymer solution and give rise to phenomena such as shear-rate dependent viscosities and non-vanishing normal-stress differences [11, 16, 17, 21, 24, 25].

The characterization of the tumbling dynamics is an important step toward a microscopic understanding of the polymer dynamics, an issue which accordingly received considerable attention [2–4, 6–8, 10, 11, 18, 26–28]. As it turns out, this is a non-trivial task because of the non-Markovian nature of the process [4, 8, 11, 29]. As is commonly accepted by now, tumbling is a cyclic but non-periodic process, which poses challenges for the calculation of a characteristic time.

The complex interactions in polymer solutions hamper an analytical treatment. Here, computer simulations are essential to shed light on their rich and intricate dynamical behavior. We apply a hybrid mesoscale hydrodynamics simulation approach, which combines molecular dynamics simulations (MD) for the polymers with the multiparticle collision dynamics (MPC) method describing the solvent [22, 23, 30–36]. As has been shown, the MPC method is very well suited to study the non-equilibrium properties of polymers [12, 37–40], colloids [13, 41, 42], and other soft-matter object such as vesicles [43] and cells [44, 45] in flow fields. Moreover, simulation studies of the dynamics of polymers in dilute solution yield excellent agreement with predictions of the Zimm model [21, 33–35].

By this approach, we demonstrate that polymers in dilute and semidilute solutions exhibit large deformations and a strong alignment with the flow direction in simple shear flow [17, 18]. More importantly, in the stationary state, the *conformational* and *rheological properties* for various concentrations are universal functions of the Weissenberg number $\mathrm{Wi_c} = \dot\gamma\tau(c)$, where $\dot\gamma$ is the shear rate and $\tau(c)$ the concentration-dependent polymer end-to-end vector relaxation time at equilibrium. Hence, with increasing concentration, hydrodynamic interactions affect the conformational and rheological properties only via the increasing relaxation time $\tau(c)$. Experiments on DNA in shear flow [20] and simulations of polymer brushes [46] lead to a similar conclusion.

By analyzing the *non-equilibrium dynamical properties*—orientational distribution functions and tumbling times, which are extracted from gyration tensor correlation functions along the flow and gradient direction—of semidilute polymer solutions, we find that the dynamical behavior depends on concentration (in excess of $\tau(c)$), a dependence which we attribute to screening of hydrodynamic interactions in semidilute solution [18]. Compared to the dilute case, such a screening causes a broadening of orientational angle distribution functions and an increase of the tumbling frequency $f = \tau(c)/\tau_T$, where τ_T is the tumbling time, at the same Weissenberg number $\mathrm{Wi_c}$ in dilute and semidilute solutions. More importantly, the same asymptotic dependencies are obtained as function of the Weissenberg number $\mathrm{Wi_c}$ in dilute and semidilute solutions. This explains the previously obtained agreement between power spectral densities of free-draining and non-draining computer simulations [2].

In this article, we briefly review the results presented in Refs. [17, 18] for polymer conformational, rheological, and dynamical properties under shear flow of dilute and semidilute solutions, and discuss their connection in a coherent manner. In Sec. 2, the model and simulation approach are described. The conformational properties are discussed Sec. 3. Results for the rheological behavior are presented in Sec. 4. Sec. 5 addresses the dynamical aspects of polymers, including the tumbling dynamics, and Sec. 6 summarizes our findings.

2. Model and Parameters

We consider a solution of N_p linear flexible polymer chains embedded in an explicit solvent. Each polymer is comprised of N_m beads of mass M, which are connected by harmonic springs. The bond potential is

$$U_b = \frac{\kappa}{2}\sum_{i=1}^{N_m-1}\left(|\boldsymbol{R}_{i+1} - \boldsymbol{R}_i| - l\right)^2, \tag{1}$$

where $\boldsymbol{R}_i$ denotes the position of monomer i, l is the bond length, and κ the spring constant. Inter- and intramolecular excluded-volume interactions are taken into account by the repulsive, shifted, and truncated Lennard-Jones potential

$$U_{LJ} = 4\epsilon\left[\left(\frac{\sigma}{r}\right)^{12} - \left(\frac{\sigma}{r}\right)^{6} + \frac{1}{4}\right], \tag{2}$$

for $r < \sqrt[6]{2}\sigma$, where r is the distance between two monomers, and $U_{LJ} = 0$ for $r \geq \sqrt[6]{2}\sigma$. Here, ϵ characterizes the energy of the interaction potential and σ the diameter of a monomer. Newton's equations of motion for the polymer are integrated by the velocity Verlet algorithm with the time step h_p [47, 48].

The solvent is simulated by the multiparticle collision (MPC) dynamics method [22, 23, 30, 31]. It is composed of N_s point-like particles of mass m, which interact with each other by a stochastic process. The algorithm consists of alternating streaming and collision steps. In the streaming step, the particles move ballistically and their positions $\boldsymbol{r}_i$ are updated according to

$$\boldsymbol{r}_i(t+h) = \boldsymbol{r}_i(t) + h\boldsymbol{v}_i(t), \tag{3}$$

where $i = 1, \ldots, N_s$, h is the time interval between collisions—denoted as collision time—, and $\boldsymbol{v}_i$ is the velocity of particle i. In the collision steps, the particles are sorted into cubic cells of side length a and their relative velocities, with respect to the center-of-mass velocity

$$\boldsymbol{v}_{cm} = \frac{1}{N_c}\sum_{j=1}^{N_c} \boldsymbol{v}_j, \tag{4}$$

of the cell of particle i, are rotated around a randomly oriented axis by a fixed angle α, i.e.,

$$\boldsymbol{v}_i(t+h) = \boldsymbol{v}_i(t) + (\mathbf{R}(\alpha) - \mathbf{E})(\boldsymbol{v}_i(t) - \boldsymbol{v}_{cm}(t)), \tag{5}$$

where $\mathbf{R}(\alpha)$ is the rotation matrix, $\mathbf{E}$ is the unit matrix, and N_c is the total number of fluid particles in that cell.

The solvent-polymer coupling is achieved by taking the monomers into account in the collision step, i.e., for collision cells containing monomers, the center-of-mass velocity reads

$$\boldsymbol{v}_{cm}(t) = \frac{\sum_{i=1}^{N_c} m\boldsymbol{v}_i(t) + \sum_{j=1}^{N_c^m} M\boldsymbol{V}_j(t)}{mN_c + MN_c^m}, \tag{6}$$

where N_c^m is the number of monomers within the considered collision cell. To ensure Galilean invariance, a random shift is performed at every collision step [49]. The collision step is a stochastic process, where mass, momentum and energy are conserved, which leads to the build up of correlations between the particles and gives rise to hydrodynamic interactions.

To impose a shear flow, we apply Lees-Edwards boundary conditions [47, 50] for systems with short chains with $N_m = 50$. A local Maxwellian thermostat is used, which we denote as Maxwell-Boltzmann Scaling (MBS), to maintain the temperature of the fluid at the desired value [51]. A parallel MPC algorithm is exploited for systems of long chains, which is based on a three-dimensional domain-decomposition approach, where particles are sorted onto processors according to their spatial coordinates [17, 52]. Here, shear flow is imposed by the opposite movement of two confining walls. The walls are parallel to the xy-plane and periodic boundary conditions are applied in the x- and y-directions. The equations of motion of the solvent particles are modified by the wall interaction [50]. We impose no-slip boundary conditions by the bounce-back rule, i.e., the velocity of a fluid particle is reverted—in the reference moving with the wall—when it hits a wall, and phantom particles in a wall are taken into account. The same rule is applied for monomers when colliding with a wall [23].

We employ the parameters $\alpha = 130°$, $h/\sqrt{ma^2/(k_BT)} = 0.1$ (k_B is Boltzmann's constant and T is temperature), $\langle N_c \rangle = 10$, $M = m\langle N_c \rangle$, $l = \sigma = a$, $k_BT/\epsilon = 1$, $h/h_p = 50$, and the bond spring constant $\kappa = 5 \times 10^3 k_BT/a^2$. The latter choice ensures that the mean of the bond length changes by less than 0.5% and the variance of the bond length distribution by 3% only, even at the largest shear rate. Transport properties of the solvent depend on h, α, $\langle N_c \rangle$ [23, 50]. Tuning these variables allows us to attain solvents with a high Schmidt number and low Reynolds number, where momentum transport dominates over mass transport. The above parameters correspond to the solvent viscosity $\eta = 8.7\sqrt{mk_BT/a^4}$ and a Schmidt number $Sc = 17$. The polymer lengths $N_m = 50$ and 250 are considered in the concentration ranges $c/c^* = 0.16 - 2.08$ and $0.17 - 10.38$, respectively, and for the shear rates $\dot{\gamma}/\sqrt{k_BT/(ma^2)} = 10^{-4} - 3 \times 10^{-1}$ and $\dot{\gamma}/\sqrt{k_BT/(ma^2)} = 10^{-6} - 3 \times 10^{-2}$, respectively. The corresponding overlap concentrations are $c^* = 0.098l^3$ and $c^* = 0.029l^3$, determined by their radii of gyration. In dilute solution, the equilibrium end-to-end vector relaxation times are $\tau_0/\sqrt{ma^2/(k_BT)} = 6169$ and 78330 [17]. Brownian MPC simulations [23, 53]—where hydrodynamic correlations are switched off—yield an approximately five times larger relaxation time than hydrodynamic MPC for $N_m = 50$.

A posteriori, knowing the non-equilibrium behavior of the polymer, it can be demonstrated that the Reynolds number Re for the polymer is smaller than unity. The radius of gyration tensor of a polymer chain is defined as

$$G_{\beta\beta'} = \frac{1}{N_m}\sum_{i=1}^{N_m}\langle \Delta r_{i,\beta}\Delta r_{i,\beta'}\rangle, \tag{7}$$

where $\Delta r_{i,\beta}$ is the position of monomer i in the center-of-mass reference frame of the polymer ($\beta, \beta' \in \{x, y, z\}$) [17]. Using its $y-$component G_{yy} as characteristic length scale, the Reynolds number Re is given by $Re = v\sqrt{G_{yy}}/\nu$, where $\nu = \eta/(m\langle N_c\rangle)$ is the kinematic viscosity of our MPC solvent. Shear implies a rotation of the fluid and the polymer with the frequency $\omega_R = \dot{\gamma}/2$ at low shear rates. Hence, we set $v = \omega_R\sqrt{G_{yy}} = \mathrm{Wi_c}\sqrt{G_{yy}}/(2\tau(c))$, which yields

$$Re = \frac{\mathrm{Wi_c} G_{yy}}{2\tau(c)\nu}. \tag{8}$$

At low Weissenberg numbers $\mathrm{Wi_c} \lesssim 10$, the conformations of a polymer are little perturbed compared to the equilibrium state, hence, $G_{yy} \approx R_{g0}^2$, where R_{g0}^2 is equilibrium radius of gyration in dilute solution. For the polymer lengths $N_m = 50$ and $N_m = 250$, $R_{g0}^2 = 24.5l^2$ and $R_{g0}^2 = 163.5l^2$, respectively, as documented in Ref. [17]. With the above relaxation times τ_0, this yields $Re < 0.1$ for $\mathrm{Wi_c} < 50$. For larger Weissenberg numbers, we take into account the deformation of the radius of gyration tensor. As shown in Ref. [17], G_{yy} decreases for $\mathrm{Wi_c} > 50$ approximately as $G_{yy} \sim 1/\sqrt{\mathrm{Wi_c}}$. With $G_{yy} \approx R_{g0}^2/\sqrt{\mathrm{Wi_c}}$, we find $Re < 0.2$ up to $\mathrm{Wi_c} = 10^4$. The applied approximations certainly overestimate the Reynolds number, since the frequency ω_R also depends on the Weissenberg number for $\mathrm{Wi_c} > 1$ and increases slower than linear with $\mathrm{Wi_c}$, as has been shown for star polymers in Refs. [13, 53].

3. Conformations

We characterized the flow strength by the Weissenberg number $\mathrm{Wi_c} = \dot{\gamma}\tau(c)$, i.e., by the concentration dependent relaxation time. In the weak-shear flow regime, with $\mathrm{Wi_c} \ll 1$, the chains are able to undergo conformational changes before the local strain has changed by a detectable amount, while in the strong shear-flow regime $\mathrm{Wi_c} \gg 1$, the chains are driven by the flow and they are not able to relax back to the equilibrium conformation. This is illustrated in Fig. 1, which displays snapshots for various flow rates. At small $\mathrm{Wi_c}$, the polymers are only weakly perturbed and are close to their equilibrium conformations, whereas large $\mathrm{Wi_c}$ imply large deformations and a strong alignment with flow. The question is, to what extent the influence of concentration on the polymer dynamics can be accounted for by a concentration-dependent Weissenberg number. As we will see, this concept applies well for all structural and rheological properties.

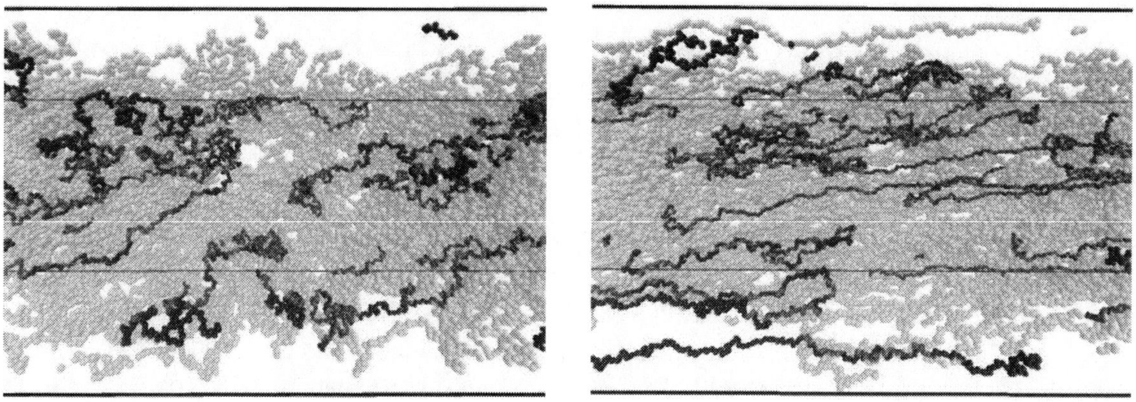

Figure 1. Snapshots of systems with $N_p = 400$ polymers of length $N_m = 250$ for the Weissenberg numbers $\text{Wi}_c = 13$ (left) and $\text{Wi}_c = 126$ (right). For illustration, some of the chains are highlighted in red.

The polymer deformation is quantified by the gyration tensor (7). In Fig. 2, the relative deformation along the flow direction

$$\delta G_{xx} = \frac{G_{xx} - G_{xx}^0}{G_{xx}^0}, \tag{9}$$

where $G_{xx}^0 = R_g^2/3$ is the gyration tensor component and R_g^2 the radius of gyration at equilibrium for the particular concentration, is shown for various concentrations and polymer lengths. A significant polymer stretching appears for $\text{Wi}_c > 1$. At large shear rates, the stretching saturates at a value smaller than the value corresponding to a fully stretched chain ($G_{xx} \approx l^2 N_m^2/12$) and reflects the finite size of a polymer. This is consistent with experiments on DNA [2, 9], where the maximum extension is on the order of half of the contour length, and theoretical calculations [16]. It is caused by the large conformational changes of the polymers during the tumbling cycles, which yields an average extension smaller than the contour length. Nevertheless, molecules assume totally stretched conformations at large Weissenberg numbers during their tumbling dynamics. Interestingly, a universal dependence is obtained for δG_{xx} as function of a concentration-dependent Weissenberg number Wi_c at a given polymer length. In contrast, polymers at larger concentrations exhibit a stronger stretching in terms of the Weissenberg number $\text{Wi} = \dot{\gamma}\tau_0$ for a dilute solution, because their actual concentration dependent relaxation time is larger and hence their Weissenberg number is higher.

Theoretical calculations for single polymers in dilute solution predict the dependence $\delta G_{xx} = C_x \text{Wi}^2$ for $\text{Wi} < 1$, where C_x is a universal constant. The renormalization group calculations of Ref. [15] yield $C_x = 0.27$, whereas a calculation based on a Gaussian phantom chain model yields $C_x \approx 0.3$ [11, 16, 54–56]. As shown in Fig. 2, the simulations confirm the quadratic dependence on the shear rate; δG_{xx} is independent of chain length for $\text{Wi} < 1$ and $C_x \approx 0.1$. For $\text{Wi} > 10$, finite-size effects appear and different asymptotic values are assumed for the two chain lengths. We like to stress that our simulations are in agreement with the molecular dynamics simulation results of Ref. [56] and the SANS data of Refs. [57, 58].

As is evident from the average shape of an individual chain displayed in Fig. 3, a polymer is not only stretched by the flow, but also aligned. This alignment is characterized by the angle χ_G, which is the angle between the eigenvector of the gyration tensor with the largest eigenvalue

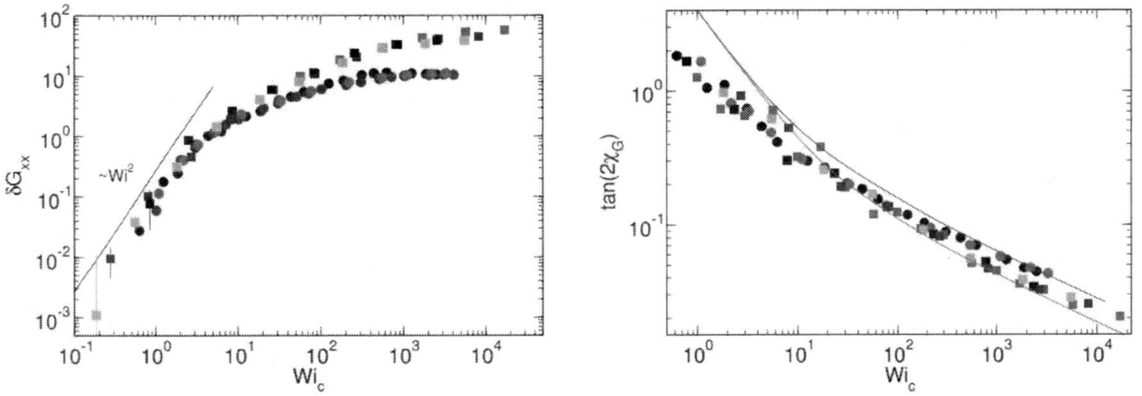

Figure 2. Deformation ratios δG_{xx} (left) and $\tan(2\chi_G)$ (right) as function of Weissenberg number. Bullets correspond to systems with $N_m = 50$ for $c/c^* = 0.16$ (●), $c/c^* = 1.6$ (●), and $c/c^* = 2.08$ (●). Squares denote results for $N_m = 250$ with the concentrations $c/c^* = 0.17$ (■), $c/c^* = 2.77$ (■), $c/c^* = 5.19$ (■), and $c/c^* = 10.38$ (■). The line in δG_{xx} indicates the dependence $\delta G_{xx} \sim \text{Wi}^2$, and the lines for $\tan(2\chi_G)$ are theoretical results according to Refs. [11, 16] for $N_m = 50$ (blue) and $N_m = 250$ (red).

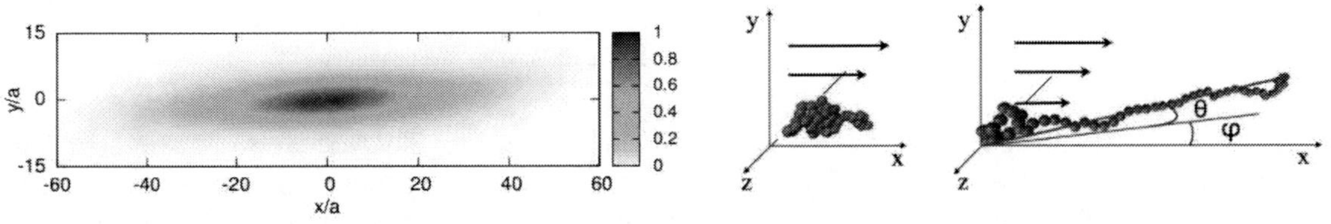

Figure 3. Monomer density distribution in the flow-gradient plane for $N_m = 250$, $c/c^* = 10.38$, and $\text{Wi}_c = 569$ (left). Illustration of polymer stretching and recoiling and the definition of the angles θ and φ (right). θ is the angle between the end-to-end vector and its projection onto the flow-gradient plane and φ is the angle between this projection and the flow direction.

and the flow direction. It is obtained from the components of the radius of gyration tensor via

$$\tan(2\chi_G) = \frac{2G_{xy}}{G_{xx} - G_{yy}}. \tag{10}$$

The dependence of $\tan(2\chi_G)$ on shear rate and concentration is shown in Fig. 2. Again, a universal curve is obtained for the different concentrations at a given polymer length. Moreover, $\tan(2\chi_G)$ seems to be independent of polymer length for $\text{Wi}_c < 100$, whereas we find a length dependence for larger Weissenberg numbers. In this high shear-rate regime, we find $\tan(2\chi_G) \sim (\text{Wi}_c)^{-1/3}$. We like to emphasize that only the shear rate can be scaled in order to arrive at a universal function. The angle, or $\tan(2\chi_G)$, cannot be scaled to absorb flow or polymer properties in an effective variable. Hence, the universal behavior of the alignment angle for various concentrations confirms that the Weissenberg number Wi_c is the correct scaling variable and that the alignment of polymers at different concentrations depends on the combination $\text{Wi}_c = \dot{\gamma}\tau_c$ of shear rate and relaxation time only.

The analytical approach of Refs. [11, 16] predicts the dependence

$$\tan(2\chi_G) \sim \left(\frac{l_p}{L\mathrm{Wi}^*}\right)^{1/3} \tag{11}$$

for semiflexible polymers in dilute solution in the limit $\mathrm{Wi}^* \to \infty$. Here, we introduce the Weissenberg number $\mathrm{Wi}^* = \dot{\gamma}\tau_{\mathrm{th}}$ for the theoretical result, because the relaxation times from theory and simulation might not be exactly the same; L is the length and l_p the persistence length of the polymer. The analytical result describes the simulation data well at large shear rates, when the Weissenberg number of the theoretical model is set to $\mathrm{Wi}^* = \mathrm{Wi}_c/2$. To compare the predicted length dependence with that of the simulation, we apply the relation $\langle R_e^2\rangle = 2l_pL$ to obtain a persistence length, with $\langle R_e^2\rangle$ the polymer mean square end-to-end distance in dilute solution at equilibrium, which yields $l_p/L \approx 0.025$ for $N_m = 50$ and $l_p/L \approx 0.008$ for $N_m = 250$. With these values, the ratio of $\tan(2\chi_G)$ of the polymer of length $L = 50a$ and $L = 250a$ is 1.5. This compares well with the factor 1.33 following from the simulation results, which suggests that excluded-volume interactions are of minor importance for intermediate flow rates.

In the limit $\mathrm{Wi}_c \to 0$, theory predicts $\tan(2\chi_G) \sim \mathrm{Wi}_c^{-1}$. The simulation data do not show this dependence on the considered range of Weissenberg numbers, which might be due to excluded-volume interactions.

4. Rheology

Applying a shear flow, the viscosity $\eta(\dot{\gamma})$ is obtained from the relation

$$\eta(\dot{\gamma}) = \frac{\sigma_{xy}(\dot{\gamma})}{\dot{\gamma}}, \tag{12}$$

where σ_{xy} is the shear stress [24, 59]. In our simulations, σ_{xy} is calculated using the virial formulation of the stress tensor [47, 50, 60]. For sufficiently weak flow, the polymer solution is in the Newtonian regime, i.e., $\sigma_{xy} \sim \dot{\gamma}$ and the viscosity is independent of shear rate. Thus, the viscosity obtained in this low shear rate regime is equal to the zero-shear viscosity denoted by η_0. A discussion of concentration dependence of η_0 can be found Ref. [17].

At sufficiently large shear rates, the polymers are aligned and deformed, which implies shear thinning [11, 14, 16, 24]. Figure 4 shows the polymer contribution η^p to the shear viscosity. Similar to the alignment angle, the viscosity is a universal function of the Weissenberg number Wi_c and shows a weak dependence on polymer length. It is independent of shear rate for $\mathrm{Wi}_c \ll 1$, decrease approximately as $\mathrm{Wi}_c^{-0.3}$ for $1 < \mathrm{Wi}_c < 10^2$, and $\mathrm{Wi}_c^{-0.45}$ for higher shear rates. This behavior is consistent with other simulation results [9, 46, 61–63]. However, an even stronger decay of the viscosity is observed in simulations at larger shear rates in Refs. [9, 14]. Experiments of dilute polymer solutions reported exponents ranging from -0.4 to -0.85 [9, 24]. Theoretical calculations for dumbbells and finite extensible polymers predict the dependence $\eta_p \sim \mathrm{Wi}^{-2/3}$ in the limit $\mathrm{Wi} \to \infty$ [11, 16, 24, 25, 64]. The differences in the observed behavior can be explained by a broad crossover regime before the asymptotic behavior is reached, as suggested by the simulations of Ref. [9]

The ratio of the viscosities of the two lengths is approximately 1.33 for the large Weissenberg-number regime, as for the alignment angle, which compares well with the theoretically predicted length dependence in Eq. (11).

The concentration and shear-rate dependencies of the first normal-stress coefficient [24, 25]

$$\Psi_1 = (\sigma_{xx} - \sigma_{yy})/\dot{\gamma}^2 \tag{13}$$

is displayed in Fig. 4. Within the accuracy of the simulations, the ratio Ψ_1/Ψ_1^0, where Ψ_1^0 is the stress coefficient at zero shear rate, is a universal function of Wi_c for various concentrations and

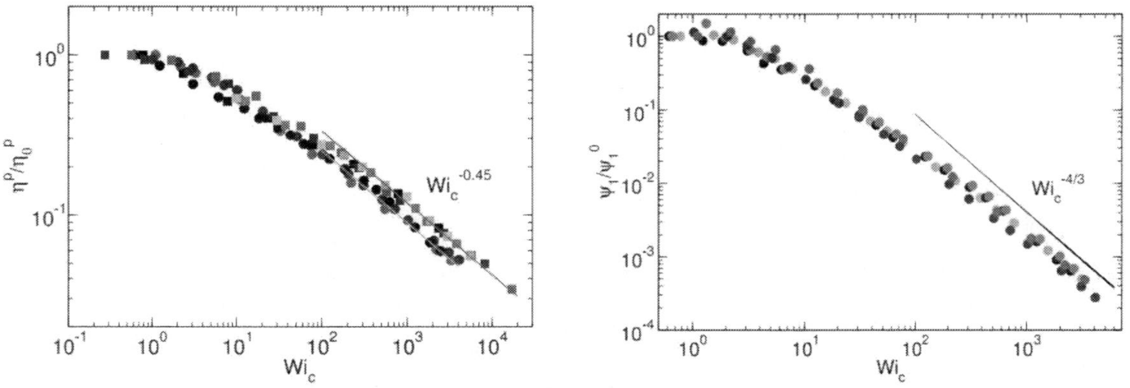

Figure 4. Polymer contribution η_p to the shear viscosity (left) and first normal-stress coefficient Ψ_1 (right) as function of shear rate. Bullets correspond to systems with $N_m = 50$ for $c/c^* = 0.16$ (●), $c/c^* = 0.41$ (●), $c/c^* = 0.81$ (●), $c/c^* = 1.63$ (●) and $c/c^* = 2.08$ (●). Squares denote results for $N_m = 250$ with the concentration $c/c^* = 0.17$ (■), $c/c^* = 0.35$ (■), $c/c^* = 1.38$ (■), $c/c^* = 2.77$ (■), $c/c^* = 5.19$ (■), and $c/c^* = 10.38$ (■).

decreases as $\Psi_1 \sim \dot{\gamma}^{-4/3}$ for large shear rates. This is consistent with analytical calculations [16, 24, 25], various computer simulations [9, 17, 25, 61–63, 65], and experiments [2, 3] for dilute solutions. Similar to the viscosity, the decay is related to the finite polymer extensibility.

Surprisingly, the first normal-stress coefficient follows the theoretical prediction [16, 24, 25] already for moderate Weissenberg numbers, whereas the viscosity η^p does not. This is consistently found in various simulation studies applying different models and simulation approaches [9, 17, 25, 61–63, 65]. The origin of the disparity is not clear. Theoretical models based on phantom polymers, i.e., models neglecting excluded-volume interactions, suggest an intimate coupling of the shear-rate dependence of the viscosity and the normal-stress coefficient [16]. It seems that excluded-volume interactions, which lead to a tight coupling of the polymer deformation along the flow and gradient direction, as expressed in the correlation function presented in Fig. 6 (left), affect the shear stress σ_{xy} more severely and lead to the observed differences. Such a coupling is not captured in a phantom chain model. This speculation could be checked by simulations of phantom chains.

5. Angular Probability Distribution Functions

The probability distribution function (PDF) $P(\varphi)$ for the orientation angle φ (for a definition of φ, see Fig. 3 (right)) provides further insight into the orientational behavior of polymers [4, 6, 8, 11]. Theoretical calculations yield the expression [11]

$$P(\varphi) = \frac{\sqrt{\pi^4 + 4\mathrm{Wi}^{*2}(\pi^2-4)/\mu^2}}{2\pi\left[\pi^2+4\right]\mathrm{Wi}^{*2}\sin^2(\varphi)/\mu^2 - 4\mathrm{Wi}^*\sin(2\varphi)/\mu} \tag{14}$$

in the limit of long and flexible polymers $L/l_p \gg 1$, where μ follows from the equation

$$\mu^3 - \mu^{5/2} - \frac{\pi^4 l_p}{270L}\mathrm{Wi}^{*2} = 0, \tag{15}$$

which yields $\mu = \mathrm{Wi}^{*2/3}\sqrt[3]{\pi^4 l_p/(270L)}$ in the asymptotic limit $\mathrm{Wi}^* \to \infty$.

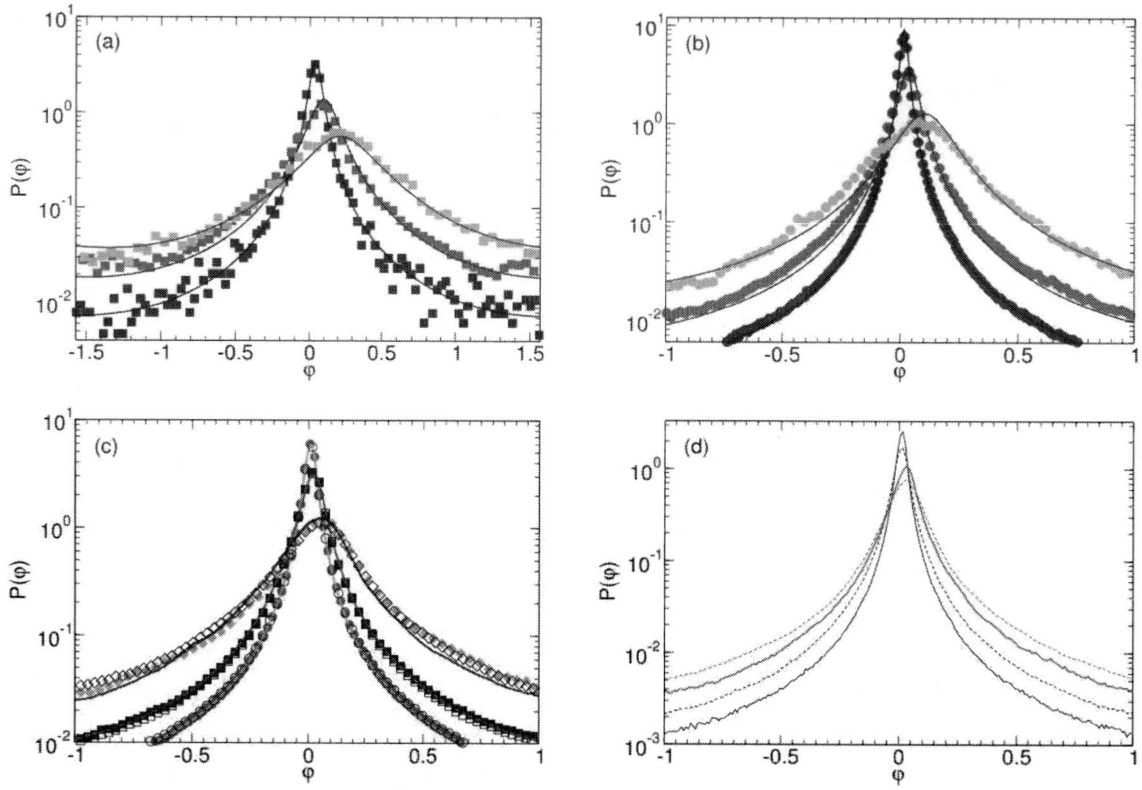

Figure 5. Probability distribution functions of the angle φ. (a) Dilute solution with $c/c^* = 0.16$ and $N_m = 50$ for $\mathrm{Wi_c} = 616.9$ (■), $\mathrm{Wi_c} = 61.7$ (■), and $\mathrm{Wi_c} = 12.3$ (■). (b) Dilute solution for $N_m = 250$, $c/c^* = 0.17$, and $\mathrm{Wi_c} = 2350$ (●), $\mathrm{Wi_c} = 235$ (●), $\mathrm{Wi_c} = 23.5$ (●). The solid lines in (a) and (b) are calculated according to Eq. (14) for the Weissenberg number $\mathrm{Wi}^* = \mathrm{Wi_c}/2$. (c) Semidilute solutions of polymers of length $N_m = 250$ for $c/c^* = 2.77$ with $\mathrm{Wi_c} = 5520$ (●), $\mathrm{Wi_c} = 552$ (■), and $\mathrm{Wi_c} = 55.2$ (◆), for $c/c^* = 5.19$ with $\mathrm{Wi_c} = 5423$ (○), $\mathrm{Wi_c} = 542.3$ (□), and $\mathrm{Wi_c} = 54.23$ (◇), as well as $c/c^* = 10.38$ with $\mathrm{Wi_c} = 5691$ (green line), $\mathrm{Wi_c} = 569.1$ (red line), $\mathrm{Wi_c} = 56.91$ (blue line). (d) Comparison of $P(\varphi)$ for dilute $c/c^* = 0.35$ (solid) and semidilute $c/c^* = 5.19$ (dashed) solutions with $\mathrm{Wi_c} = 2670$ (black solid line), $\mathrm{Wi_c} = 267$ (red solid line), $\mathrm{Wi_c} = 2700$ (black dashed line), and $\mathrm{Wi_c} = 270$ (red dashed line).

For a *dilute* solution, $P(\varphi)$ is shown in Figs. 5(a) and (b) for various Weissenberg numbers, together with theoretical lines obtained from Eq. (14) [11] (note that $\mathrm{Wi}^* = \mathrm{Wi_c}/2$). Evidently, the simulation results agree well with the analytical approach, as is expected for a dilute solution in which the intermolecular interactions are irrelevant. $P(\varphi)$ exhibits a significant shear-rate dependence. Without any shear, no angle is preferred. A finite shear rate leads to the appearance of a peak, which shifts to smaller values with increasing $\dot{\gamma}$ and, at the same time, the width $\Delta\varphi$ of $P(\varphi)$ decreases. Figure 5(c) displays distribution functions for various concentrations and Weissenberg numbers $\mathrm{Wi_c}$ of semidilute solutions. For every Weissenberg number, distributions are compared for three concentrations $c > c^*$. Evidently, the distributions are almost independent of concentration for the considered Weissenberg numbers. However, we observe a clear concentration dependence, when we compare distributions from dilute and semidilute solutions. Figure 5(d), displays distributions for the concentrations $c/c^* = 0.35$ and 5.19 and the Weissenberg numbers $\mathrm{Wi_c} \approx 267$, 2670 and 270, 2700, respectively. The increase in concentration from a dilute solution beyond the overlap concentration leads to a broadening of the distribution function. Interestingly, the value φ_m at the peak of the distribution function is independent of concentration at a given $\mathrm{Wi_c}$ [18]. Theoretical calculations show that φ_m is close to the alignment angle χ_G [11].

6. Tumbling Dynamics

As has been shown [2, 18], the tumbling dynamics of a polymer can be characterized by and a characteristic time be obtained from the cross-correlation function

$$C_{xy}(t) = \frac{\langle G'_{xx}(t_0) G'_{yy}(t_0 + t)\rangle}{\sqrt{\langle G'^2_{xx}(t_0)\rangle \langle G'^2_{yy}(t_0)\rangle}}, \tag{16}$$

for deviations from average stationary values $G'_{\beta\beta}(t) = G_{\beta\beta}(t) - \langle G_{\beta\beta}\rangle$ of the radius of gyration tensor components along the flow and gradient direction. Examples of cross-correlation functions are displayed in Fig. 6 for several shear rates and concentrations. Every curves exhibits a deep minimum at a time $t_+ > 0$ and a pronounced maximum at a time $t_- < 0$, and decays to zero at large time-lags. Hence, the tumbling dynamics is not periodic, but cyclic. The latter has been questioned for tethered polymers [26]. The minimum at t_+ indicates that positive values of $G'_{\beta\beta}$ are linked with negative ones of the orthogonal directions, i.e., polymer shrinkage in the y-direction is linked with its extension in x-direction, and similarly, an extension in y-direction is linked to shrinkage in x-direction. The maximum of $C_{xy}(t)$ reveals that positive deviations G'_{xx} are correlated with positive values G'_{yy} at earlier times, or a collapsed state along the x-direction ($G'_{xx} < 0$) is correlated with a previous collapsed state in y-direction [2]. Hence, the time difference $t_+ - t_-$ is related to conformational changes that a polymer undergoes during tumbling, which is characterized by the time $\tau_t = 2(t_+ - t_-)$ [2]. The factor two accounts for the fact that two non-equivalent conformations lead to a maximum and a minimum, respectively, and will be (more or less) assumed during a cycle.

As shown in Fig. 6, the positions of the maxima and minima are rather close for equal Weissenberg numbers, when the lag-time is scaled by the relaxation time $\tau(c)$. Hence, the tumbling times exhibit a strong concentration dependence due to the concentration dependence of the relaxation times [17].

Normalized tumbling frequencies $f = \tau(c)/\tau_T$, with tumbling times extracted from the correlation functions and scaled by the corresponding relaxation times $\tau(c)$, are presented in Fig. 6 (right) for a wide range of shear rates and concentrations. For comparison, the theoretical prediction for a polymer in dilute solution is presented as well [11, 16]. The results agree very well. The short chain results clearly show the crossover from unity, assumed in the limit $\dot{\gamma} \to 0$, to the asymptotic dependence $\sim \mathrm{Wi_c^{2/3}}$ at high shear rates. We obtain a chain-length dependence in close agreement with the theoretical prediction. More importantly, we find a slight and gradual shift of f to larger values with increasing concentration at a given $\mathrm{Wi_c}$, until a saturation is reached in the semidilute regime $c/c^* > 1$. This is seen for the two largest concentrations for $N_m = 50$ and the three largest ones for $N_m = 250$. As a consequence, the polymers exhibit a universal behavior, both in dilute ($c \ll c^*$) as well as in semidilute solution as function of $\mathrm{Wi_c}$, with the same power-law dependence on $\mathrm{Wi_c}$ (for $\mathrm{Wi_c} > 1$).

Alternatively, relaxation times under shear flow can be obtained by the end-to-end vector auto-correlation function $\langle R_\beta(t) R_\beta(0)\rangle$, where $R_\beta = R_{N_m,\beta} - R_{1,\beta}$ [10, 11, 28, 66, 67]. Similar to the results presented in Refs. [28, 66], we find a damped oscillatory behavior for $\mathrm{Wi_c} \gtrsim 1$ [67]. By fitting the obtained correlation function with the function

$$F_\beta(t) = a_\beta e^{-t/\tau_\beta^{(1)}} [\cos(\omega_\beta t) + b_\beta \sin(\omega_\beta t)] + [1 - a_\beta] e^{t/\tau_\beta^{(2)}}, \tag{17}$$

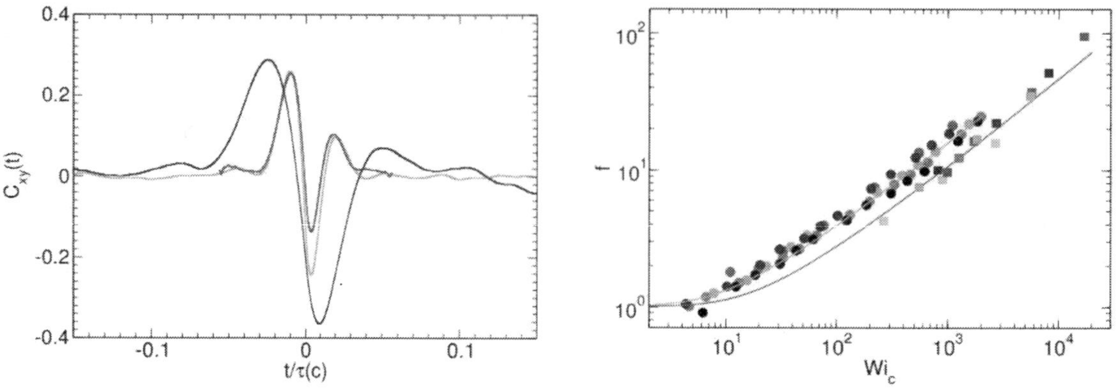

Figure 6. Left: Cross-correlation functions [Eq. (16)] for a polymer of length $N_m = 250$ and the concentrations $c/c^* = 2.77$ (green), $c/c^* = 10.38$, (red), and $c/c^* = 0.35$ (blue), corresponding to the Weissenberg numbers $\mathrm{Wi_c} = 5520$, $\mathrm{Wi_c} = 5690$, and $\mathrm{Wi_c} = 2670$, respectively. Right: Normalized tumbling frequencies $f = \tau(c)/\tau_T$. Bullets correspond to the polymer length $N_m = 50$ for $c/c^* = 0.16$ (●), $c/c^* = 0.41$ (●), $c/c^* = 0.81$ (●), $c/c^* = 1.63$ (●), and $c/c^* = 2.08$ (●). Squares indicate results for $N_m = 250$ and $c/c^* = 0.35$ (■), $c/c^* = 0.69$ (■), $c/c^* = 1.38$ (■), $c/c^* = 2.77$ (■), $c/c^* = 5.19$ (■), $c/c^* = 10.38$ (■). The lines present the theoretical predictions [11, 16].

with the amplitudes a_β, b_β, relaxation times $\tau_\beta^{(i)}$, and frequencies ω_β, we obtain non-equilibrium relaxation times. Strictly speaking, the functional form of the correlation function is unknown. As expressed by the correlation function C_{xy} (16), tumbling is non-periodic and, hence, Eq. (17) applies over a certain time window only. We expect the equation to capture correlations in the polymer conformations over a time scale comparable to a tumbling cycle.

By fitting, we obtain the following general relations for the various shear rates and densities: $a_x \approx a_z \approx 1$, $a_y \approx 1/2$, and $\omega_z = 0$, $\omega_x \approx \omega_y$. Since $a_y \approx 1/2$, the decay in the gradient direction is governed by two exponentials, where $\tau_y^{(2)}$ is an order of magnitude smaller than $\tau_y^{(1)}$, but shows a similar shear-rate dependence. Since $\omega_z = 0$, the correlation function along the vorticity direction decays exponentially. Moreover, b_x and b_y, where b_y is negative, are approximately independent of shear rate, but depend on polymer length. Most importantly, the relaxation times $\tau_x^{(1)}$, $\tau_y^{(1)}$, and $\tau_z^{(1)}$ are equal in dilute solution and agree with the tumbling times displayed in Fig. 6 (right). Hence, the polymer tumbling times are equal to the non-equilibrium end-to-end relaxation times. The parameter ω_β is independent of shear rate for the considered systems. This is in contrast to results of Ref. [66], where very short polymers ($N_m = 10$) have been considered only.

We attribute the concentration independence of the tumbling time and the probability distribution functions for $c/c^* > 1$ to screening of hydrodynamic interactions. To confirm our hypothesis, we performed Brownian MPC simulations for dilute and semidilute solutions [18]. Using similar Weissenberg numbers, we find, within the accuracy of the simulations, identical distribution functions $P(\varphi)$ for both cases. Moreover, the distributions agree with those of semidilute systems of the same concentration and Weissenberg number in the presence of hydrodynamic interactions. Hence, the differences between distribution functions at low and high concentrations, as displayed in Fig. 5(d), are due to hydrodynamic interactions. In dilute solutions, hydrodynamic interactions are present, whereas in systems with $c > c^*$, hydrodynamic interactions are screened. Naturally, at larger concentrations friction is higher. This aspect is captured in the relaxation time $\tau(c)$, which increases considerably with concentration [17].

The fact that the tumbling frequencies and the widths of the distribution functions are larger in semidilute solutions, i.e., when hydrodynamic interactions are screened, might be explained as follows. (i) The observed broadening of the distribution function $P(\varphi)$ in semidilute solution implies that hydrodynamic interactions favor polymer alignment and lead to a faster dynamics during the collapse and stretching part of the tumbling motion. (ii) At the same Weissenberg number, the shear rate of a non-draining polymer is larger than that of a free-draining one, due to differences in equilibrium relaxation times, i.e., for a coiled conformation. As a consequence, the effective Weissenberg number $\mathrm{Wi}_R = \dot{\gamma}\tau_R$, where τ_R is the rotational relaxation time in the stretched rodlike conformation, of the non-draining polymer is larger than that of the free-draining one. This could explain the larger probability of angles in the vicinity of φ_m for non-draining polymers as well as their faster collapse dynamics. Overall, the tumbling time is larger in a non-draining system.

The broadening of the distribution functions with increasing concentration or screening of hydrodynamic interactions is not captured by standard theories employing the preaveraging approximation [11, 16, 21]. Here, hydrodynamic interactions are included in the relaxation times and hence the Weissenberg number only; additional, "higher order effects" are neglected. Therefore, one might expect that the theoretical description would reproduce results of simulations without hydrodynamic interactions. In contrast, the model calculations rather reproduce the simulation data for systems with hydrodynamic interactions.

7. Conclusions

We have calculated conformational, dynamical, and rheological properties of polymers in dilute and semidilute solution under shear flow by mesoscale hydrodynamic simulations. We find that their stationary-state conformational and rheological properties are independent of concentration when expressed in terms of the Weissenberg number $\mathrm{Wi}_c = \dot{\gamma}\tau(c)$. This is remarkable, since the longest polymer relaxation time $\tau(c)$ increases significantly with concentration and indicates that an effective local friction determines the stationary-state properties.

By analyzing dynamical properties—orientational distribution functions and tumbling times—of semidilute polymer solutions, we find that they depend on concentration (in excess of $\tau(c)$), a dependence which we attribute to screening of hydrodynamic interactions in semidilute solution. Compared to the dilute case, such a screening causes a broadening of orientational angle distribution functions and an increasing ratio $f = \tau(c)/\tau_T$ at the same Weissenberg number Wi_c in semidilute solution. The effect itself is small ($f(c=0)/f(c>c^*) \approx 1.3$ at $\mathrm{Wi}_c = 10^3$). More importantly, the same asymptotic dependencies are obtained as function of the Weissenberg number Wi_c in dilute and semidilute solutions. This explains the previously obtained agreement of power spectral densities obtained from free-draining and non-draining computer simulations [2].

Moreover, we find that the tumbling times almost quantitatively agree with the non-equilibrium end-to-end vector relaxation times, which, hence, exhibit the asymptotic shear rate dependence $\sim \dot{\gamma}^{-2/3}$ for $\dot{\gamma} \to \infty$.

Our simulations reveal a complex interplay between shear rate, deformation, and intramolecular excluded-volume interactions, which is difficult to grasp by analytical theory.

Acknowledgments

Financial support by the Deutsche Forschungsgemeinschaft within SFB TR6 "Physics of Colloidal Dispersions in External Fields" is gratefully acknowledged. We are grateful to the Jülich Supercomputer Centre (JSC) for allocation of a special CPU-time grant.

References

[1] Smith D E, Babcock H P and Chu S 1999 *Science* **283** 1724
[2] Schroeder C M, Teixeira R E, Shaqfeh E S G and Chu S 2005 *Phys. Rev. Lett.* **95** 018301
[3] Teixeira R E, Babcock H P, Shaqfeh E S G and Chu S 2005 *Macromolecules* **38** 581
[4] Gerashchenko S and Steinberg V 2006 *Phys. Rev. Lett.* **96** 038304
[5] Doyle P S, Ladoux B and Viovy J L 2000 *Phys. Rev. Lett.* **84** 4769
[6] Celani A, Puliafito A and Turitsyn K 2005 *Europhys. Lett.* **70** 464
[7] Chertkov M, Kolokolov I, Lebedev A and Turitsyn K 2005 *J. Fluid. Mech.* **531** 251
[8] Puliafito A and Turitsyn K 2005 *Physica D* **211** 9
[9] Schroeder C M, Teixeira R E, Shaqfeh E S G and Chu S 2005 *Macromolecules* **38** 1967
[10] Delgado-Buscalioni R 2006 *Phys. Rev. Lett.* **96** 088303
[11] Winkler R G 2006 *Phys. Rev. Lett.* **97** 128301
[12] Winkler R G, Mussawisade K, Ripoll M and Gompper G 2004 *J. Phys.: Condens. Matter* **16** S3941–S3954
[13] Ripoll M, Winkler R G and Gompper G 2006 *Phys. Rev. Lett.* **96** 188302
[14] Aust C, Kröger M and Hess S 1999 *Macromolecules* **32** 5660
[15] Wang S Q 1990 *J. Chem. Phys.* **92** 7618
[16] Winkler R G 2010 *J. Chem. Phys.* **133** 164905
[17] Huang C C, Winkler R G, Sutmann G and Gompper G 2010 *Macromolecules* **43** 10107
[18] Huang C C, Sutmann G, Gompper G and Winkler R G 2011 *EPL* **93** 54004
[19] Babcock H P, Smith D E, Hur J S, Shaqfeh E S G and Chu S 2000 *Phys. Rev. Lett.* **85** 2018
[20] Hur J, Shaqfeh E S G, Babcock H P, Smith D E and Chu S 2001 *J. Rheol.* **45** 421
[21] Doi M and Edwards S F 1986 *The Theory of Polymer Dynamics* (Oxford: Clarendon Press)
[22] Kapral R 2008 *Adv. Chem. Phys.* **140** 89
[23] Gompper G, Ihle T, Kroll D M and Winkler R G 2009 *Adv. Polym. Sci.* **221** 1
[24] Bird R B, Hassager O, Armstrong R C and Curtiss C F 1987 *Dynamics of Polymeric Liquids* vol 2 (New York: John Wiley Sons)
[25] Öttinger H C 1996 *Stochastoic Processes in Polymeric Fluids* (Berlin: Springer)
[26] Zhang Y, Donev A, Weisgraber T, Alder B J, Graham M G and de Pablo J J 2009 *J. Chem. Phys.* **130** 234902
[27] Kobayashi H and Yamamoto R 2010 *Phys. Rev. E* **81** 041807
[28] Usabiaga F B and Delgado-Buscalioni R 2011 *Macromol. Theory Simul.* **20** 466
[29] Das D and Sabhapandit S 2008 *Phys. Rev. Lett.* **101** 188301
[30] Malevanets A and Kapral R 1999 *J. Chem. Phys.* **110** 8605
[31] Malevanets A and Kapral R 2000 *J. Chem. Phys.* **112** 7260–7269
[32] Malevanets A and Yeomans J M 2000 *Europhys. Lett.* **52** 231–237
[33] Ripoll M, Mussawisade K, Winkler R G and Gompper G 2004 *Europhys. Lett.* **68** 106
[34] Ripoll M, Mussawisade K, Winkler R G and Gompper G 2005 *Phys. Rev. E* **72** 016701
[35] Mussawisade K, Ripoll M, Winkler R G and Gompper G 2005 *J. Chem. Phys.* **123** 144905
[36] Padding J T and Louis A A 2006 *Phys. Rev. E* **73** 031402
[37] Ryder J F and Yeomans J M 2006 *J. Chem. Phys.* **125** 194906
[38] Cannavacciuolo L, Winkler R G and Gompper G 2008 *EPL* **83** 34007
[39] Frank S and Winkler R G 2008 *EPL* **83** 38004
[40] Chelakkot R, Winkler R G and Gompper G 2010 *EPL* **91** 14001
[41] Padding J T and Louis A A 2004 *Phys. Rev. Lett.* **93** 220601
[42] Wysocki A, Royall C P, Winkler R G, Gompper G, Tanaka H, van Blaaderen A and Löwen H 2009 *Soft Matter* **5** 1340
[43] Noguchi H and Gompper G 2004 *Phys. Rev. Lett.* **93** 258102
[44] Noguchi H and Gompper G 2005 *Proc. Natl. Acad. Sci. USA* **102** 14159–14164
[45] McWhirter J L, Noguchi H and Gompper G 2009 *Proc. Natl. Acad. Sci. USA* **106** 6039
[46] Galuschko A, Spirin L, Kreer T, Johner A, Pastorino C, Wittmer J and Baschnagel J 2010 *Langmuir* **26** 6418
[47] Allen M P and Tildesley D J 1987 *Computer Simulation of Liquids* (Oxford: Clarendon Press)
[48] Swope W C, Andersen H C, Berens P H and Wilson K R 1982 *J. Chem. Phys.* **76** 637
[49] Ihle T and Kroll D M 2001 *Phys. Rev. E* **63** 020201(R)
[50] Winkler R G and Huang C C 2009 *J. Chem. Phys.* **130** 074907
[51] Huang C C, Chatterji A, Sutmann G, Gompper G and Winkler R G 2010 *J. Comput. Phys.* **229** 168
[52] Sutmann G, Huang C C, Winkler R G and Gompper G 2010 *John von Neumann Institute for Computing NIC Symposium 2010* (*IAS Series* vol 3) ed Münster G, Wolf D and Kremer M (Jülich: Forschungszentrum Jülich) pp 287–294

[53] Ripoll M, Winkler R G and Gompper G 2007 *Eur. Phys. J. E* **23** 349
[54] Bruns W and Carl W 1993 *Macromolecules* **26** 557
[55] Carl W and Bruns W 1994 *Macromol. Theory Simul.* **3** 295
[56] Pierleoni C and Ryckaert J P 1995 *Macromolecules* **28** 5097
[57] Lindner P and Oberthur R C 1988 *Colloid Polym. Sci.* **266** 886
[58] Lindner P and Oberthur R C 1989 *Physica B* **156 & 157** 410
[59] Bird R B, Armstrong R C and Hassager O 1976 *Dynamics of Polymeric Liquids* vol 1 (New York: Wiley)
[60] Winkler R G, Morawitz M and Yoon D Y 1992 *Mol. Phys.* **75** 669
[61] Liu T W 1989 *J. Chem. Phys.* **90** 5826
[62] Petera D and Muthukumar M 1999 *J. Chem. Phys.* **111** 7614
[63] Hsieh C C and Larson R G 2004 *J. Rheol.* **48** 995
[64] Rubinstein M and Colby R 2003 *Polymer Physics* (Oxford: Oxford University Press)
[65] Zylka W 1991 *J. Chem. Phys.* **94** 4628
[66] Jose P P and Szamel G 2008 *J. Chem. Phys.* **128** 224910
[67] Haung C C, Gompper G and Winkler R G 20012 submitted for publication

Intrinsic viscosity for Brownian particles of arbitrary shape

Bogdan Cichocki[1], **Maria L Ekiel-Jeżewska**[2] **and Eligiusz Wajnryb**[2]

[1] Institute of Theoretical Physics, Faculty of Physics, University of Warsaw, Hoża 69, 00-681 Warsaw, Poland

[2] Institute of Fundamental Technological Research, Polish Academy of Sciences, Pawińskiego 5B, 02-106 Warsaw, Poland

E-mail: mekiel@ippt.gov.pl

Abstract. General expressions for the frequency-dependent Brownian contribution to the intrinsic viscosity of arbitrary-shaped particles have been derived from the Smoluchowski equation.

Dedicated to François Feuillebois on the occasion of his 65th birthday

1. Introduction

In physical chemistry, dynamic light scattering and viscosity measurements can provide information about structure of particles (such as e.g. fibrinogen) in colloidal suspensions [1]. Indeed, a change of the particle shape modifies its self-diffusion, and in general leads to a significant change of its intrinsic viscosity. For suspensions of arbitrary-shaped particles, the single-particle short-time self-diffusion coefficient and the high-frequency intrinsic viscosity can be easily derived from the existing theoretical framework, e.g. with the use of the bead model, and the multipole expansion of the Stokes equations [2]. However, as far as we know, the Brownian contribution to the intrinsic viscosity has been determined only for special types of shapes, such as axially symmetric particles [3, 4]. In this work, we theoretically derive general expressions for the frequency-dependent Brownian contribution to the intrinsic viscosity of arbitrary-shaped particles.

2. Problem

We start from reminding the basic equation, which describes motion of a single particle of arbitrary shape in the Stokes flow [5]. Assuming that there exist external force and torque $\mathbf{F}, \mathbf{T}$ exerted on the particle, and an ambient linear flow $\mathbf{v}_0$, we express the particle translational and rotational velocities, $\mathbf{U}$ and $\tilde{\mathbf{\Omega}}$, and the stresslet $\mathbf{S}$, by the following relation [5],

$$\begin{bmatrix} \mathbf{U} - \mathbf{v}_0(\boldsymbol{R}) \\ \tilde{\mathbf{\Omega}} - \boldsymbol{\omega}_0(\boldsymbol{R}) \\ -\mathbf{S} \end{bmatrix} = \begin{bmatrix} \boldsymbol{\mu}^{tt} & \boldsymbol{\mu}^{tr} & \boldsymbol{\mu}^{td} \\ \boldsymbol{\mu}^{rt} & \boldsymbol{\mu}^{rr} & \boldsymbol{\mu}^{rd} \\ \boldsymbol{\mu}^{dt} & \boldsymbol{\mu}^{dr} & \boldsymbol{\mu}^{dd} \end{bmatrix} \cdot \begin{bmatrix} \mathbf{F} \\ \mathbf{T} \\ \mathbf{g}_0 \end{bmatrix}, \tag{1}$$

where $\boldsymbol{R}$ is the particle center, $\boldsymbol{\omega}_0 = \mathrm{rot}\, \mathbf{v}_0/2$ and $\mathbf{g}_0$ is the rate-of-strain tensor. The grand mobility matrix in the above equation consists of single-particle mobility tensors $\boldsymbol{\mu}$ with the corresponding upper indices, as indicated above.

We consider a spatially uniform suspension of the particle concentration $n = N/V$ with N being the number of suspended particles in volume V. The suspension is subjected to an oscillating shear flow $\mathbf{v}_0$ with the rate of strain

$$\mathbf{g}_0(t) = \mathbf{g}_{0\omega} \exp(-i\omega t). \tag{2}$$

Other quantities oscillate with the same frequency, and their amplitudes are denoted in a similar way, i.e. by adding the lower index ω.

In the presence of particles, the rate of strain (2) is modified. Introducing $\langle ... \rangle$ as the average with respect to the ensemble of particle positions and orientations, we write the amplitude of the effective rate of strain as $\langle \mathbf{g} \rangle_\omega$. For small-amplitude oscillations, the effective viscous part of the stress tensor amplitude $\langle \boldsymbol{\sigma} \rangle_\omega^v$ and the effective rate of strain amplitude $\langle \mathbf{g} \rangle_\omega$ are linearly related, with the coefficient equal to the effective viscosity $\eta_{\text{eff}}(\omega)$,

$$\langle \boldsymbol{\sigma} \rangle_\omega^v = 2\eta_{\text{eff}}(\omega) \langle \mathbf{g} \rangle_\omega . \tag{3}$$

Taking into account the results of the Chapter 2.5 from Ref. [5] one can write

$$\langle \boldsymbol{\sigma} \rangle_\omega^v = 2\eta_0 \langle \mathbf{g} \rangle_\omega - \langle \mathbf{S} \rangle_\omega \frac{N}{V}, \tag{4}$$

where $-\langle \mathbf{S} \rangle_\omega$ is the amplitude of average particle stresslet oscillations and η_0 is the solvent viscosity.

We consider low particle concentrations N/V, corresponding to low volume fractions $\phi = Nv/V$, where v is the particle volume, with no hydrodynamic interactions between the particles. In this case, the suspension effective viscosity can be written as

$$\eta_{\text{eff}}(\omega) = \eta_0 \left(1 + [\eta]_\omega \phi + ...\right), \tag{5}$$

where the coefficient $[\eta]_\omega$ is called the intrinsic viscosity. Value of the intrinsic viscosity depends on a particle shape and its internal structure. In particular, for a hard sphere, $[\eta]_\omega = 5/2$.According to Eq. (5), we are interested to determine the linear dependence of the effective viscosity on volume fraction. Within this approximation, $\langle \mathbf{g} \rangle_\omega$ in Eqs. (3)-(4) can be replaced by $\mathbf{g}_{0\omega}$.

The intrinsic viscosity consists of two parts: the high-frequency contribution $[\eta]_\infty$ and the Brownian contribution $[\eta]_\omega^B$,

$$[\eta]_\omega = [\eta]_\infty + [\eta]_\omega^B . \tag{6}$$

The high-frequency contribution $[\eta]_\infty$ is obtained from Eqs. (3)-(4) as the average of $\mathbf{S}$ with respect to the equilibrium probability distribution; for an isotropic system, $\langle \mathbf{S} \rangle_\omega$ is proportional to $\mathbf{g}_{0\omega}$, with a scalar proportionality coefficient evaluated by the double contraction of both tensors. Therefore, according to Eq. (1),

$$[\eta]_\infty = \frac{\mu_{kllk}^{dd}}{10\eta_0 v}, \tag{7}$$

where $k, l = x, y, z$ denote the Cartesian indices of the matrix $\boldsymbol{\mu}^{dd}$, and the Einstein summation convention has been used.

3. The Smoluchowski equation

To evaluate $[\eta]_\omega^B$, we first need to introduce the probability distribution and the Smoluchowski equation. For a spatially uniform suspension, only the rotational motion contributes to the Brownian part of the effective viscosity $[\eta]_\omega^B$ [6]. The orientation of an anisotropic particle is given by the Euler angles $\widehat{\mathbf{\Omega}} = (\alpha, \beta, \gamma)$ which specify the orientation of the particle body-fixed axes with respect to the space-fixed axes. We use the notation proposed by Berne and Pecora in Ref. [7]. We denote the orientational probability distribution at time t by $P(\widehat{\mathbf{\Omega}}, t)$. It is normalized as

$$\int d\widehat{\mathbf{\Omega}} P(\widehat{\mathbf{\Omega}}, t) = 1, \tag{8}$$

with

$$\int d\widehat{\mathbf{\Omega}} \ldots = \int_0^{2\pi} d\alpha \int_0^{2\pi} d\gamma \int_0^{\pi} d\beta \sin\beta \ldots \tag{9}$$

Evolution of the probability distribution $P(\widehat{\mathbf{\Omega}}, t)$ is governed by the Smoluchowski equation,

$$\frac{\partial}{\partial t} P(\widehat{\mathbf{\Omega}}, t) = \mathcal{L}(\widehat{\mathbf{\Omega}}, t) P(\widehat{\mathbf{\Omega}}, t). \tag{10}$$

In the absence of an ambient flow, $\mathcal{L}(t) = \mathcal{L}_0$, with

$$\mathcal{L}_0 = -\mathbf{I} \cdot \mathbf{D}^r \cdot \mathbf{I} \tag{11}$$

where

$$\mathbf{D}^r = k_B T \, \boldsymbol{\mu}^{rr} \tag{12}$$

is the tensor of the rotational diffusion, and

$$\mathbf{I} = (I_x, I_y, I_z) \tag{13}$$

is the total angular momentum operator with components being the angular momentum operators about axes x, y, z stuck in the particles. If α_k is the angle of rotation around axis k then

$$I_k = -i \frac{\partial}{\partial \alpha_k}. \tag{14}$$

In the presence of an external flow,

$$\mathcal{L}(t) = \mathcal{L}_0 + \delta\mathcal{L}(t). \tag{15}$$

When the particle is located in a rate-of-strain field characterized by $\mathbf{g}_0(t)$,

$$\delta\mathcal{L}(t) = -i\mathbf{I} \cdot \boldsymbol{\mu}^{rd} \cdot \mathbf{g}_0(t), \tag{16}$$

where the contraction of $\boldsymbol{\mu}^{rd}$ and the strain tensor $\mathbf{g}_0(t)$ is performed with respect to two Cartesian indices.

4. Rotational Brownian contributions to the intrinsic viscosity

To find the Brownian contribution $[\eta]_\omega^B$ to the intrinsic viscosity (6), we must calculate the time-dependent averages, which appear in Eqs. (3)-(4). These averages are performed only over the angles,

$$\langle \ldots \rangle_t = \int d\widehat{\mathbf{\Omega}} \, P(\widehat{\mathbf{\Omega}}, t) \ldots. \tag{17}$$

Within the linear response theory [8, 9], the solution of Eq. (10), with $\mathcal{L}(t)$ given by Eqs. (15)-(16), has the following form,

$$P(\widehat{\mathbf{\Omega}},t) = P_{eq}(\widehat{\mathbf{\Omega}}) + \int_{-\infty}^{t} dt' \exp[\mathcal{L}_0(t-t')]\delta\mathcal{L}(t')P_{eq}(\widehat{\mathbf{\Omega}}). \quad (18)$$

The contribution to the non-Brownian intrinsic viscosity, coming from averaging with the equilibrium distribution, P_{eq}, has been already described in Eq. (7). The Brownian contribution follows entirely from the averaging with respect to the non-equilibrium part of P.

In case of the Brownian rotation, the average stresslet is due to the Brownian torque,

$$-\langle \mathbf{S}^B \rangle_t = \langle \boldsymbol{\mu}^{dr} \cdot \mathbf{T}_{Br} \rangle_t, \quad (19)$$

with [5]

$$\mathbf{T}_{Br} = -ik_BT\ \mathbf{I} \ln P(\widehat{\mathbf{\Omega}},t). \quad (20)$$

The expression (19) may be transformed to

$$-\langle \mathbf{S}^B \rangle_t = k_BT\ \langle \mathbf{H} \rangle_t, \quad (21)$$

with

$$H_{ij} = iI_k \mu^{rd}_{kij}. \quad (22)$$

To obtain the above equation, we performed integration by parts in angles and took into account the antisymmetric property of the generalized mobility matrix elements with dr and rd indices, i.e.

$$\mu^{dr}_{ijk} = -\mu^{rd}_{kij}. \quad (23)$$

The symmetric and traceless tensor $\mathbf{H}$ can be derived from $\boldsymbol{\mu}^{rd}$ by the irreducible tensor analysis described in Ref. [10],

$$H_{ij} = \epsilon_{ikl}\mu^{rd}_{lkj} + \epsilon_{jkl}\mu^{rd}_{lki}. \quad (24)$$

Let us now consider an oscillating shear flow with the rate of strain, see Eq. (2), and assume that the amplitude $\mathbf{g}_{0\omega}$ is small. From Eq. (16) we calculate $\delta\mathcal{L}(t)$ and insert it into Eq. (18). Then, we obtain the following relation between the oscillation amplitudes $\langle \mathbf{S}^B \rangle_\omega$ and $\mathbf{g}_{0\omega}$,

$$-\langle S^B_{ij} \rangle_\omega = \left[\frac{k_BT}{8\pi^2}\int d\widehat{\mathbf{\Omega}}\ H_{ij}(\widehat{\mathbf{\Omega}}) \frac{1}{i\omega + \mathbf{I}\cdot\mathbf{D}^r\cdot\mathbf{I}} H_{kl}(\widehat{\mathbf{\Omega}})\right] g_{0\omega,kl}, \quad (25)$$

where the Einstein summation convention has been used.

For the isotropic system, $\langle S^B_{ij} \rangle_\omega$ is proportional to $g_{0\omega,ij}$. It is convenient to write the scalar proportionality coefficient using the double contraction of both tensors. Taking into account Eqs. (3)-(4), we therefore write,

$$[\eta]^B_\omega = \frac{k_BT}{80\pi^2 v\eta_0} \int d\widehat{\mathbf{\Omega}}\ H_{ij}(\widehat{\mathbf{\Omega}}) \frac{1}{i\omega + \mathbf{I}\cdot\mathbf{D}^r\cdot\mathbf{I}} H_{ij}(\widehat{\mathbf{\Omega}}), \quad (26)$$

with contraction of the repeated indices.

To find the inverse operator in Eq. (26), we now choose the body-fixed frame in which the rotational diffusion tensor $\mathbf{D}^r$ is diagonal,

$$\mathbf{D}^r_{ij} = D_i\delta_{ij}, \quad \text{for } i,j = x,y,z. \quad (27)$$

In this frame, we consider the eigen-problem for the operator

$$\mathbf{I}\cdot\mathbf{D}^r\cdot\mathbf{I}=(D_xI_x^2+D_yI_y^2+D_zI_z^2), \tag{28}$$

and construct the eigen-solution [7, 11] as

$$(D_xI_x^2+D_yI_y^2+D_zI_z^2)\psi^{(J)}_{K,M}(\widehat{\mathbf{\Omega}})=f^{(J)}_K\ \psi^{(J)}_{K,M}(\widehat{\mathbf{\Omega}}),$$
$$\text{for } J=0,1,2,\dots \text{ and } K,M=-J,-J+1,\dots,J \tag{29}$$

The tensor $\mathbf{H}$ is symmetric and traceless; therefore, to invert the operator in Eq. (26), we need explicit expressions for $\psi^{(J)}_{K,M}$ and $f^{(J)}_K$ with $J=2$ only. The functions $\psi^{(2)}_{K,M}$ form a complete orthonormal set in our space of the considered functions. In Ref. [7], the eigen-functions are given in terms of the normalized Wigner functions $C^{(2)}_{K,M}$,

$$\psi^{(2)}_{K,M}(\widehat{\mathbf{\Omega}}) = \sum_{m'} a_{K,m'}C^{(2)}_{m',M}(\widehat{\mathbf{\Omega}}). \tag{30}$$

The coefficients $a_{K,m'}$ are listed in table 1. They depend only on the rotational diffusion

Table 1. The coefficients $a_{K,m'}$.

$K\setminus m'$	-2	-1	0	1	2
-2	$-\frac{1}{\sqrt{2}}$	0	$-\frac{a}{\sqrt{2(a^2+b^2)}}$	0	$\frac{b}{\sqrt{2(a^2+b^2)}}$
-1	0	$-\frac{1}{\sqrt{2}}$	0	$\frac{1}{\sqrt{2}}$	0
0	0	0	$\frac{b}{\sqrt{a^2+b^2}}$	0	$\frac{a}{\sqrt{a^2+b^2}}$
1	0	$\frac{1}{\sqrt{2}}$	0	$\frac{1}{\sqrt{2}}$	0
2	$\frac{1}{\sqrt{2}}$	0	$-\frac{a}{\sqrt{2(a^2+b^2)}}$	0	$\frac{b}{\sqrt{2(a^2+b^2)}}$

coefficients, and are expressed in terms of the combinations,

$$a = \sqrt{3}(D_x-D_y), \tag{31}$$
$$b = 3(D_z-D)+2\Delta, \tag{32}$$

with

$$D=\frac{1}{3}\left(D_x+D_y+D_z\right). \tag{33}$$

In Eq. (29), the eigen-values with $J=2$ are given as,

$$f^{(2)}_2 = 6D+2\Delta, \tag{34}$$
$$f^{(2)}_0 = 6D-2\Delta, \tag{35}$$
$$f^{(2)}_1 = 3\left(D_x+D\right), \tag{36}$$
$$f^{(2)}_{-1} = 3\left(D_y+D\right), \tag{37}$$
$$f^{(2)}_{-2} = 3\left(D_z+D\right), \tag{38}$$

where

$$\Delta = \left(D_x^2 + D_y^2 + D_z^2 - D_xD_y - D_xD_z - D_yD_z\right)^{1/2}. \tag{39}$$

In Eq. (38), the modes are listed taking into account that they split into two groups: the modes with the indices $K = 2, 0$ are similar to each other, and the same holds for the modes with $K = 1, -1, -2$. Therefore, we introduce new labels, and we denote $f_i = f_K^{(2)}$, with $i = 1, 2$ corresponding to $K = 2, 0$, and $i = 3, 4, 5$ corresponding to $K = 1, -1, -2$, respectively.

Taking into account the completeness of the eigenfunctions $\psi_{K,M}^{(2)}(\widehat{\mathbf{\Omega}})$, and using Appendix of Ref. [7], we obtain from Eq. (26) a simple expression for the Brownian intrinsic viscosity,

$$[\eta]_\omega^B = \frac{k_BT}{10v\eta_0}\sum_{i=1}^{5}\frac{A_i}{i\omega + f_i}, \tag{40}$$

where $A_i = A_K^{(2)}$, with $i = 1, 2, 3, 4, 5$ corresponding to $K = 2, 0, 1, -1, -2$, respectively, and

$$A_K^{(2)} = \sum_{M=-2}^{2}\iint\frac{d\widehat{\mathbf{\Omega}}\,d\widehat{\mathbf{\Omega}}'}{8\pi^2}H_{ij}(\widehat{\mathbf{\Omega}})\psi_{K,M}^{(2)*}(\widehat{\mathbf{\Omega}})H_{ij}(\widehat{\mathbf{\Omega}}')\psi_{K,M}^{(2)}(\widehat{\mathbf{\Omega}}'). \tag{41}$$

To evaluate the above integrals, we express ψ in terms of the Wigner functions (see Eq. (30)), and then apply the expressions from Appendix of Ref. [7] for the rotational integrals of symmetric traceless tensors projected onto Wigner functions. Using this method, we calculate the coefficients A_i in Eq. (40). For $i = 3, 4, 5$, they depend only on the rotational-dipole mobility $\boldsymbol{\mu}^{rd}$,

$$A_3 = 2H_{yz}^2, \tag{42}$$
$$A_4 = 2H_{zx}^2, \tag{43}$$
$$A_5 = 2H_{xy}^2, \tag{44}$$

with the symmetric traceless tensor $H_{\alpha\beta}$ given by Eq. (24). Explicitly,

$$H_{yz} = \mu_{xzz}^{rd} - \mu_{zxz}^{rd} + \mu_{yxy}^{rd} - \mu_{xyy}^{rd}, \tag{45}$$
$$H_{zx} = \mu_{yxx}^{rd} - \mu_{xyx}^{rd} + \mu_{zyz}^{rd} - \mu_{yzz}^{rd}, \tag{46}$$
$$H_{xy} = \mu_{zyy}^{rd} - \mu_{yzy}^{rd} + \mu_{xzx}^{rd} - \mu_{zxx}^{rd}. \tag{47}$$

The coefficients A_i in Eq. (40), with $i = 1, 2$, depend on the rotational-dipole mobility and the rotational diffusion coefficients,

$$A_1 + A_2 = H_{xx}^2 + H_{yy}^2 + H_{zz}^2, \tag{48}$$
$$A_1 - A_2 = \frac{1}{\Delta}\left[\bar{c}_xD_x + \bar{c}_yD_y + \bar{c}_zD_z\right], \tag{49}$$

where

$$\bar{c}_x = -2H_{xx}^2 + H_{yy}^2 + H_{zz}^2, \tag{50}$$
$$\bar{c}_y = H_{xx}^2 - 2H_{yy}^2 + H_{zz}^2, \tag{51}$$
$$\bar{c}_z = H_{xx}^2 + H_{yy}^2 - 2H_{zz}^2, \tag{52}$$

and from Eq. (24),

$$H_{xx} = 2(\mu_{zyx}^{rd} - \mu_{yzx}^{rd}), \tag{53}$$
$$H_{yy} = 2(\mu_{xzy}^{rd} - \mu_{zxy}^{rd}), \tag{54}$$
$$H_{zz} = 2(\mu_{yxz}^{rd} - \mu_{xyz}^{rd}). \tag{55}$$

5. Conclusions

In this work, we have theoretically derived general frequency-dependent expressions for the Brownian contribution to the intrinsic viscosity of arbitrary-shaped particles, see Eq. (40). The intrinsic viscosity depends on single-particle rotational-rotational and rotational-dipole mobilities. The effect of Brownian rotations on dilute suspensions of rigid particles of arbitrary shape was studied in Ref. [12] by another method (rotation-matrix formalism), but with no explicit expressions for the intrinsic viscosity.

In the present work, performing the time-dependent analysis, we have evaluated the characteristic time scales which determine when the high-frequency or zero-frequency approximations to the intrinsic viscosity are adequate. In a next publication, we will explain how to use the expressions given here to calculate explicitly the Brownian intrinsic viscosity, and will perform computations for a few particle shapes of a given symmetry, relevant for practical applications.

Acknowledgments

We thank John Hinch for helpful discussions during the symposium *Microparticles in Stokes Flows*, and informing us about the existence of Ref. [12].

References

[1] Wasilewska M, Adamczyk Z and Jachimska B 2009 *Langmuir* **25** 3698
[2] Cichocki B, Felderhof B U, Hinsen K, Wajnryb E, and Bławzdziewicz J 1994 *J. Chem. Phys.* **100** 3780
[3] Brenner H 1974 *J. Multiphase Flow* **1** 195
[4] Wajnryb E and Dahler J S 1997 *Adv. Chem. Phys.* **102** 193
[5] Kim S and Karrila S J 2005 *Microhydrodynamics. Principles and Selected Applications.* (Mineola: Dover Publications)
[6] Cichocki B, Ekiel-Jeżewska M L and Wajnryb E 2012 *J. Chem. Phys.* **136** 071102
[7] Berne B and Pecora R 1976 *Dynamic Light Scattering: With Applications to Chemistry, Biology and Physics* (New York: Wiley) chapter 7.
[8] Kubo R, Toda M and Hashitsume N 1985 *Statistical Mechanics II* (Berlin: Springer)
[9] Felderhof B U and Jones R B 1987 *Physica A* **146** 417
[10] Jerphagnon J, Chemla D and Bonneville R 1978 *Adv. Phys.* **27** 609
[11] Favro L D 1960 *Phys. Rev.* **119** 53
[12] Rallison J M 1978 *J. Fluid Mech.* **84** 237

Vesicles, capsules and red blood cells under flow

Chaouqi Misbah

Université Grenoble I/CNRS, Laboratoire Interdisciplinaire de Physique/UMR5588, Grenoble F-38041, France

E-mail: `chaouqi.misbah@ujf-grenoble.fr`

Abstract. Blood flow is dictated by the dynamics of red blood cells (RBCs), which constitute by far the major component. RBCs are made of a a two dimensional fluid bilayer of phospholipids, having underneath a network of proteins conferring to them shear elasticity, and they possess many membrane and transmembrane proteins (like ion channels). Simplified systems, like vesicles (made of a pure bilayer of phospholipid) and capsules (made of an extensible polymer shell) are used as models for RBCs. Both systems reproduce several features known for RBCs under flow. Their interest lies, besides some simplicity, in the fact that they can be fabricated in the laboratory, and their properties (size, stiffness, internal content....) can be varied in a wide range allowing thus to explore a quite significant parameter space that is essential to test predictions and discriminate between different models. We shall review the main recent achievement in this field, both for a single entity, collective effects and the impact on rheology.

1. Introduction

Blood is a complex fluid that is primarily composed of red blood cells (RBCs), which occupy (in a healthy human body), about 45% of the blood volume. The rest consists of plasma, while the other blood elements (white blood cells, platelets, etc.) take up less than 1% of the total blood volume.

The descriptions of blood flow properties escape the traditional laws for simple fluids. The complex character results from an intimate coupling between the shape of RBCs and the ambient plasma, which leads to a rich set of RBC morphologies in the blood circulatory system. Understanding the selection of shapes and dynamics among a large manifold of possibilities, the collective effects, the spatiotemporal organizations, is a challenging problem. This type of complexity is a characteristic property of non-equilibrium dissipative systems for which general thermodynamic principles, such as minimization of energy, maximization of entropy, etc., cannot be applied.

During the 70's and 80's a large number of studies were devoted to finding the equilibrium shape of a RBC. Helfrich [1] introduced simple geometrical models in which he assumed that most of the energy is stored in bending modes of the cell membrane. He proposed that the energy of the cell is a quadratic functional of the mean membrane curvature H, and the shape of the cell is obtained by minimizing the curvature energy subject to two constraints: the total area is constant (the local area, and thus the total one, is fixed –the membrane is viewed as a two dimensional incompressible fluid), and the enclosed volume is also constant, the internal fluid being incompressible. The two constraints are imposed via two Lagrange multipliers λ_1

and λ_2, so that the total "free-energy" to be minimized is

$$F = \frac{\kappa}{2}\int H^2 dA + \lambda_1 \int dA + \lambda_2 \int dV \tag{1}$$

where the first two integral are performed over the total surface, the last one over the enclosed volume, κ is the membrane bending rigidity, having the dimension of an energy. The curvature energy (the first term) is invariant under an increase of the absolute size of the cell (if lengths (x, y, z) are multiplied by a constant μ –a homothety–, this implies that $dA \to \mu^2 dA$ and $H^2 \to \mu^{-2}H^2$, so that the energy is invariant). This implies that the size itself does not matter, but only the swelling ratio, that tells us how much is the shape close to, or far from, a sphere. The swelling ratio, called also the reduced volume, is defined as the ratio of the actual volume V over that of a sphere having the same area A as the RBC:

$$\nu = \frac{V/(4\pi/3)}{[A/4\pi]^{3/2}} \tag{2}$$

For a sphere $\nu = 1$, otherwise $\nu < 1$. For a human RBC $\nu \simeq 0.65$. This model was successful in predicting the biconcave shape of the RBC, but also in predicting many other shapes corresponding to different values of ν . RBCs have been mimicked by vesicles (the membrane of which is made of a bilayer of phospholipids, unlike RBCs, which in addition, possess proteins networks and protein channels, etc..., see late) that could be fabricated in the laboratory, and where the reduced volume could be varied at will in order to test the prediction of the model. A quite good agreement could be found between predictions and experiments, provided that the model be enriched, like introducing, for example, a spontaneous curvature H_0 (where H in (1) had to be substituted by $H - H_0$, meaning that an open membrane would have minimal curvature energy when $H = H_0$). The problem acquires now two parameters, ν and H_0, and the variety of shapes increases quite dramatically. The problem of equilibrium shapes by using the Helfrich model and its variants is now well understood both experimentally and theoretically (see [2] for a review).

The end of 90's has known the emergence of a new activity, namely the behaviours of RBCs and vesicles under nonequilibrium conditions (basically under imposed flows), albeit some sparse works appeared already in the 80's without active research at the international level. While by the end 90's only few groups (basically in Europe) were active in this field, the advent of the new century has known an unbelievable upsurge of interest in various communities, physics, mechanics, mechanical engineering and applied mathematics, everywhere over the world, with a central activity in Europe, USA and Asia. The number of papers that has been devoted to this field since the end of 1990 may be estimated to few hundreds. Therefore this review can not cover all these progresses, but will focus on some aspects, namely vesicles, capsules and RBCs under shear and Poiseuille flows, and only say few rapid words about rheology of a dilute suspension, and collective effects of RBCs.

2. RBC properties and introduction of the biomimetic systems, vesicles and capsules

Healthy human RBC at rest has a biconcave shape that minimizes the Helfrich energy. Its size if of about few μm in length, and its membrane is made of a phospholipid bilayer, which is fluid at physiological temperature. Besides phospholipid molecules, the membrane possesses a spectrin network (also called the cytoskeleton) made of proteins that confer to RBC plane shear elasticity (figure 1). In addition, the membrane of RBC has several membrane and transmembrane proteins having different roles and functions (like ion channels...). For mechanical properties suffices it to say here that the RBC membrane has a visco-elastic behaviors. The interior

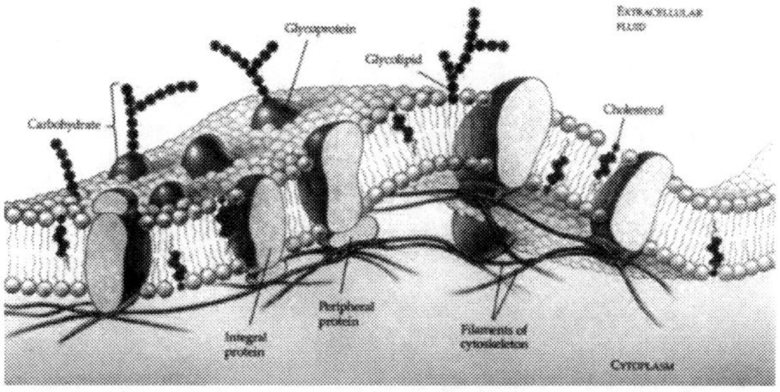

Figure 1. A schematic view of the the RBC membrane showing the phospholipid bilayer, the spectrin network (called cytoskeleton) beneath and several membrane and transmembrane proteins.

of the RBC is made of an aqueous solution containing hemoglobin (and other species, like ATP-adenosine triphosphate) that is responsible for oxygen transport from lungs towards the microvasculature (arterioles, veinules and capillaries) where gas exchange takes place. The hemoglobin solution is a newtonian fluid. The hemoglobin solution has, at physiological temperature, a viscosity which is of about 5 times that of the plasma (its viscosity is close to that of water), but this viscosity is quite temperature-dependent.

A vesicles is a closed membrane which is made, like RBC, of a purely phospholipid bilayer, which is fluid at physiological as well as at room temperatures. It has become now quite standard to fabricate them in the research laboratories, with various sizes (their typical radius ranges from few μm to about 100 μm). Their bending rigidity is of about a fraction of eV (and is close to that of the RBCs). They can be swelled or emptied thanks to osmosis, and thus their reduced volume ν can be changed. Their internal content can be varied (polymer solutions in order to act on the internal viscosity, and so on). In particular for $\nu = 0.65$ the vesicles exhibits a biconcave shape which is similar to that of a human RBC. Different types of phospholipids can also be used, in order to act, for example, on the bending rigidity. Vesicles share with RBCs an important property, namely, their membrane area is inextensible.

Finally, capsules are quasi-spherical shells made of polymers. They can be obtained thanks to interfacial polymerization of a liquid drop. This leads to approximately spherical particles enclosed by a thin polymerized membrane with mechanical properties that depend on the fabrication process. For biological applications (as carrier of active substances, for example) the typical membranes that are used or natural or synthetic polymers such as poly-L-lysine, alginate or polyacrylates. Their membrane is extensible, unlike vesicles and RBCs, in turn they are endowed with shear elasticity, mimicking the cytoskeleton of the RBCs. Their sizes can be quite diverse (say from a fraction of μm up to few mm; the later size is quite familiar in everyday life, the pharmaceutical capsule). The mechanics of a capsule belongs to the class of nonlinear elasticity [3].

3. Basic modeling of RBCs and their biomimetic counterparts under flow

In vivo the Reynolds number at the scale of the RBC is often small enough (except in arterioles and at some sites of aneurysms [4]) so that the inertia is often negligible. Experiments on vesicles on RBCs have been so far confined to the small Reynolds number limit (but access to higher Reynolds numbers is by now quite feasible). The fluid inside and outside the cell is thus described by the Stokes equations

$$\eta \Delta \mathbf{u} - \nabla P = 0 \tag{3}$$

Journal of Physics: Conference Series **392** (2012) 012005 doi:10.1088/1742-6596/392/1/012005

$$\nabla \cdot \mathbf{u} = 0 \tag{4}$$

where $\mathbf{u}$ is the velocity field and P is the pressure. The quantities associated with the internal fluid are denoted with a prime. For example, the viscosity of the internal fluid (which is in general different from that of the external fluid) will be denoted as η'. At the membrane we have the following boundary conditions:

- Continuity of the velocity field
$$\mathbf{u} = \mathbf{u}' \tag{5}$$
The continuity of the normal component is a consequence of the mass conservation, while the tangential velocity continuity follows from the assumption of no-slip condition.
- Equality of the fluid velocity and the membrane one, $\mathbf{u}_m$
$$\mathbf{u} = \mathbf{u}' = \mathbf{u}_m \tag{6}$$
This is valid under the assumption that the membrane is not permeable to water flow (which is the case once the osmotic equilibrium is achieved). Note that the knowledge of the normal membrane velocity (which coincides with that of the adjacent fluid) is sufficient in order to determine the evolution of the membrane. However, if one is interested in the fluid motion along the membrane (the so-called tank-treading motion), then we have to determine the adjacent fluid tangential velocity. To that end we impose that the tangential membrane velocity coincide with that of the adjacent fluid (the no-slip condition).
- Continuity of the stress
$$\sigma.\mathbf{n} - \sigma'.\mathbf{n} + \mathbf{f} = 0 \tag{7}$$
where $\sigma = -P\mathbf{I} + \eta(\nabla\mathbf{u} + \nabla\mathbf{u}^T)$ is the stress tensor, with the superscript 'T' designating the transpose and $\mathbf{f}$ is the membrane force. This force is composed of two contributions: the bending force and the membrane in-plane elasticity (for example membrane shear elasticity), that we shall introduce separately.

Bending force $\mathbf{f}_b$

The bending force, which can be obtained from the functional derivative of the energy (1), and $\mathbf{n}$ is the normal vector pointing towards the exterior of the cell. Before writing this expression two remarks are in order: (i) the incompressibility condition in (4) guarantees, in principle, the conservation of the enclosed volume; the Lagrange multiplier λ_2 can thus be removed from (1), the pressure field P plays now the role of the Lagrange multiplier. (ii) Under flow the stresses are inhomogeneous along the membrane, so that we need to impose local membrane incompressibility, and not only the global one. In other words, the global Lagrange multiplier must be replaced by a local one, denoted hereafter as ζ, that plays on the membrane the same role played by the pressure field in the bulk (the membrane is a two dimensional incompressible fluid). The functional derivative of the free energy provides us with the membrane force [1, 5] (a simple derivation can be found in two dimension in [6])

$$\mathbf{f}_b = \left[k_c\left(\frac{1}{2}(H - H_0)[4K - H(H + H_0)] - \Delta_s H\right) + H\zeta\right]\mathbf{n} + \nabla_s\zeta \tag{8}$$

where H is the mean curvature, K the Gaussian curvature and Δ_s is the surface Laplacian (known also as the Laplace-Beltrami operator) and they can all be written in terms of the normal vector

$$H = \nabla_s.\mathbf{n} = \nabla.\mathbf{n} \quad \text{and} \quad 2K = H^2 - \nabla\mathbf{n} : \nabla\mathbf{n}^T. \tag{9}$$

The surface gradient, the surface divergence and the Laplace-Beltrami operator are respectively expressed by:

$$\nabla_s f = (\mathbf{I} - \mathbf{n} \otimes \mathbf{n})\ \nabla f = \nabla f - (\mathbf{n}.\nabla f)\ \mathbf{n}, \quad (10)$$
$$\nabla_s.\mathbf{v} = (\mathbf{I} - \mathbf{n} \otimes \mathbf{n}) : \nabla \mathbf{v} = \nabla.\mathbf{v} - ((\nabla \mathbf{v}).\mathbf{n}).\mathbf{n}, \quad (11)$$
$$\Delta_s f = \nabla_s.(\nabla_s f)\,. \quad (12)$$

f and $\mathbf{v}$ are scalar and vector fields, respectively.

- The Lagrange multiplier ζ is determined from the requirement that the surface divergence of the velocity field is zero

$$\nabla_s.\mathbf{u} = 0 \quad (13)$$

This condition will is used when modeling a RBC or a vesicle (which are inextensible), but not for extensible capsules, of course. One can alternatively impose quasi-incompressibility by choosing a large membrane tension, as done in some numerical calculation [7]. If a cytoskeleton is to be taken into account, the associated force has to be added to the membrane bending force. Since several preliminaries are needed, this force will be introduced separately in the next section.

- Finally, boundary conditions on the external domain (containing all the suspended entities) must be specified. In an unbounded domain, the velocity field at infinity must be equal to the imposed one (e.g. linear shear flow, or Poiseuille flow...). If the domain is finite, say a tube, or a cube, usually the boundaries are taken to be rigid, and the field is assumed to vanish on the boundaries. Of course, in the human body, the blood vessels are flexible, and for those problems we have to specify the dynamics of the blood vessel wall. This question will not be considered in this short review.

Elastic force $\mathbf{f}_e$

Consider a small patch of a membrane and try to write down how that patch will act on its environment (i.e. on the membrane materials). Like in usual bulk elasticity, due to the short range interaction, the patch interact with its environment only via the contour of the patch. This means that the force must be a divergence of a second-order tensor

$$\mathbf{f}_e = \nabla_s.\tau \quad (14)$$

where τ is the stress tensor defined in the plane of the membrane [1]. It is interesting to note that also the bending force in (9) can also be written as a surface divergence of a second-order tensor τ_b, given by

$$\tau_b = -\kappa \left(-\frac{1}{2} H^2 (\mathbf{I} - \mathbf{n} \otimes \mathbf{n}) + H \nabla_s \otimes \mathbf{n} - \mathbf{n} \otimes \nabla_s H \right) + (\mathbf{I} - \mathbf{n} \otimes \mathbf{n})\zeta \quad (15)$$

It can be shown, in the theory of elasticity of large deformations, that

$$\tau = \frac{1}{J_S} \mathbf{F}^S . \frac{\partial W_s}{\partial \mathbf{E}^S} . \mathbf{F}^{S^T} \quad (16)$$

where W_s is the surface strain energy function, $\mathbf{F}^S = (\mathbf{I}-\mathbf{n}\otimes\mathbf{n}).\mathbf{F}.(\mathbf{I}-\mathbf{n}^R\otimes\mathbf{n}^R)$ with F is the usual deformation tensor (if the reference configuration is labeled as $\mathbf{x}^R$, and the actual one as $\mathbf{x}(\mathbf{x}^R, t)$,

[1] Integrating (15 on the patch of area A and using the Gauss theorem on a curved surface, one has $\int_A \nabla_s.\tau = \oint_L \tau\mathbf{m} + \int_A 2H\tau\mathbf{n}$, where L is the contour delimiting the patch, $\mathbf{m}$ is the normal to the contour. If the stress tensor τ i defined on the tangent plane, then $\tau\mathbf{n} = 0$, and only the contour contribution remains, in accord with the Cauchy principle of elasticity

$=\partial \mathbf{x}/\mathbf{x}^R$), and $\mathbf{E}^S$ is the Green-Lagrange tensor defined as $\mathbf{E}^S = \frac{1}{2}(\mathbf{F}^{S^T}.\mathbf{F}^{S^T} - \mathbf{I} + \mathbf{n}^R \otimes \mathbf{n}^R)$ (that coincides with the usual two dimensional strain tensor in the limit of linear elasticity), and $J^S = \mathrm{Det}(\mathbf{F}^S)$. In order to obtain the explicit form of the force, one needs to introduce an explicit model for the energy W_s. Like the Helfrich energy which depends on invariants (the mean curvature is nothing but the trace of the tensor $\nabla_s \mathbf{n}$), this energy must depend on invariant only. Two commonly used invariants are

$$I_1 = 2\mathrm{Trace}(\mathbf{F}^S), \quad I_2 = J_s^2 - 1 \tag{17}$$

Thus W_s depends on these two invariants, $W_s(I_1, I_2)$. This is true as long as one considers that the membrane is an isotropic medium. As stated above the elasticity of RBCs is ensured by a protein network (spectrin). This network is quite disordered. From one elementary cell of the network to another the branches of spectrin exhibits all possible orientations, and therefore one expects the isotropy assumption to make sense. The widely used model model for the RBC and capsules is that of Skalak et al. [8]

$$W_s = \frac{B}{4}[\frac{1}{2}I_1^2 + I_1 - I_2] + \frac{C}{8}I_2 \tag{18}$$

where B and C are parameter constant estimated to $B \simeq 0.005$ dyn/cm and $C \simeq 100$ dyn/cm. The large magnitude of C compared to that of B ensures that a small deviation of I_2 from unity generates large tensions. Consequently the membrane is nearly incompressible. The idea is to inject (18) into (16) in order to obtain the force (15) that has to be added inserted into (7) where $\mathbf{f} = \mathbf{f}_b + \mathbf{f}_e$. The above set of equations is complete and describe dynamics of vesicles (fluid membranes; no shear elasticity), or a RBC and a capsule if $\mathbf{f}_e$ is taken into account.

4. Different methods of solutions

There has been in the literature several methods used to solve the above set of equations. The analytical method is confined to situations where the shape is close to a sphere, and where reductive perturbation methods (transformation of the set of PDE's into nonlinear ODE's) become possible. This has been successful in studying a single vesicle [9, 10, 11, 12, 13], compressible capsule [14, 15], incompressible capsule [16, 17] under unbounded shear and Poiseuille flows. Pairs interactions has been also studied along these lines [18]. The analytical method has allowed to gain some interesting insight into the phenomena. Besides these situations which are confined to quasi-spherical shapes, all the other studies are based on numerical simulations. The first well studied method is the so-called boundary integral one, based on the use of the Green's function techniques (thanks to the linearity of the Stokes equations) [19]. This method has been used for vesicles in 2D [20, 21], and 3D [22, 7] and for capsules [23]. The advantage of this method is the fact that there is no need to solve for the fluid domain, and the whole cell dynamics is encoded in the membrane itself. This is done at a certain price, namely nonlocality: in order to move one membrane point, we must have information on the location of all the other points, so that the complexity of the method (in terms of numerical schemes) is of order N^2, N being the number of discretization points on the membranes (and boundaries, if any). Using fast multipole methods, one can lower the complexity of the method down to N [24, 25, 26].

The other class of methods consists in either meshing the fluid domains (finite differences, finite elements), or using spectral methods, and solve the fluid equations subjected to boundary conditions. Then the cell is considered to be immersed in that domain (the associated method is called the immersed boundary method). The determination of the velocity field of the membrane is performed thanks to an interpolation from neighboring lattice points where the velocity field is computed. Once this step is done, the membrane point is displaced by that velocity, and so on.

In this problem the membrane is considered in the Lagrangian sense, it is explicitly displaced. The immersed method is quite old [27, 28], but has been used only recently for studying vesicles and capsules [29, 30, 31]. In the above situations the membrane is treated in the continuum formulation, and membrane forces are then discretized. Other alternatives have treated the cell membrane as made of beads connected with springs having stretching and bending stiffness [32, 33].

Other quite popular methods that have been adapted to vesicles, and cells in general, is the phase-field [34, 35, 36, 37, 38] and the level set methods [39, 4] where the membrane is implicitly defined via the introduction of a scalar field (the phase field or the level set function), and where all geometrical properties (like mean curvature, normal, etc...) can be expressed in terms of this field. This method is fully eulerian, and has been used for vesicles, but not yet, to our knowledge, when membrane shear elasticity is included. The fluid/structure coupling is then solved either by finite differences [40], in Fourier space [35], or using the finite element methods [39, 4].

Finally, there are methods which consist not in solving the fluid equations directly, but just by mimicking them via kinetic equations. In this category, one can cite three kinds of methods (i) DPD (dissipative particle dynamics), which is akin to molecular dynamics, consists in viewing fluid particles as "atoms" or "molecules" to which basic Newton laws are applied with pairwise interactions. Despite the simplicity of the implementation method in MD, it sets a severe limitation on the size and time scales that can be handled in practice. In order to circumvent this limitation, DPD has been introduced and can be viewed as a coarse-grained method where each "particle" represents a cluster of atoms or molecules and as a result of internal degree of freedoms associated with each "particle, the particle-particle interaction includes a random and dissipative contribution [41, 42, 43]. The cell membrane is simulated by a particles interconnected with spring in the triangulated network. The particle-particle interaction are relatively soft (in comparison to real atom-atom interaction where Lennard-Jones potentials are used) and this makes the DPD method to be computationally highly efficient. However, this method might induce unacceptable errors if the degree of coarse-graining is not adequately chosen. Note also that this method, while used recently to simulate concentrated suspensions [44], has not yet been quantitatively used for the simple case of a vesicle (or RBCs) under a linear shear flow, as described in the following. (ii) Multiparticle collision dynamics (MPCD). It is a particle-based mesoscale simulation technique which incorporates thermal fluctuations and hydrodynamic interactions. Coupling of embedded particles to the coarse-grained solvent is achieved through molecular dynamics [45, 46, 47, 48, 49, 50]. This method has been used to vesicles or capsules [51, 52, 53]. This method is costly and some serious discrepancies with the boundary integral method (that uses directly the Stokes equations) have been pointed out for vesicles under a linear shear flow [7]. More precisely, in Ref.[51] it was reported that under a shear flow the period of tumbling times the shear rate varies by about a factor ten in a certain range of variation of shear rate, while in Ref. [7] only a very weak variation (of about 10%) was found. (iii) Lattice-Boltzmann method (LBM) consists in viewing the fluid as a set of particles leaving on lattice. In the spirit of the LBM, a fluid is seen as a cluster of pseudofluid particles that can collide with each other when they spread under the influence of external applied forces. Advantages of the LB method are its relative ease of implementation together with its versatile adaptability to quite arbitrary geometries. The LBM method is a hydrodynamics solver, and the cell is treated in an immersed boundary method, as discussed above [54, 55, 56, 57]. Here again, despite its use for concentrated suspensions, no quantitative phase diagram even in a simple situation (linear shear flow) has been reported yet.

Journal of Physics: Conference Series **392** (2012) 012005 doi:10.1088/1742-6596/392/1/012005

5. Type of flows and definition of dimensionless parameters

In what follows we shall review the problem of a vesicle under a linear shear flow and a planar or cylindrical Poiseuille flows. The behavior in a Couette flow will not be discussed, despite the fact that it reveals some interesting features [58]. These flows are defined as

$$v_{0x} = \dot{\gamma}y, \quad v_{0x} = v_{max}[1 - \frac{y^2}{(w/2)^2}], \quad v_{0z} = v_{max}[1 - \frac{r^2}{(w/2)^2}] \tag{19}$$

where $\dot{\gamma}$ is the shear rate, and the other velocity components are zero. These flows could be viewed as unbounded or bounded. In the latter case, and in the case of a linear shear flow we have $\dot{\gamma} = V_0/w$ where , for example, the lower plate at $y = -w/2$ is fixed (the full velocity field vanishes there), while the upper one moves with velocity V_0 (the full velocity field is equal there to V_0). For a Poiseuille flow, the typical shear rate will be taken to be that at the wall, for example in the planar Poiseuille flow it is equal to $\dot{\gamma} = 4V_{max}/w$. In the unbounded regime and for a linear shea flow $\dot{\gamma}$ is a parameter specifying the shear rate. In the case of an unbounded Poiseuille flow the absolute value of the velocity does not matter (due to Galilean invariance), but only the curvature of the flow, which is $\alpha = 8V_{max}/w^2$ (of course neither V_{max} nor w have a physical sense, but only α).

The full problem possesses several dimensionless numbers. We use R (radius of the cell corresponding to a spherical shape having the same area A as the real shape) as unit of length, $1/\dot{\gamma}$ as unit of time (in the case of a Poiseuille flow we use the shear rate at the wall, be it fictitious –unbounded flow– or real), so that the velocity unit is $U = \dot{\gamma}R$. For the pressure field we use $\eta U/R_0$. The dimensionless numbers are

- the so-called Capillary number (in analogy with drops) measuring the flow strength over the bending energy of the membrane

$$C_\kappa = \frac{\eta R^3 \dot{\gamma}}{\kappa} \equiv \dot{\gamma}\tau_\kappa \tag{20}$$

where τ_κ is the time scale of relaxation of bending modes. Typical values of κ for vesicles is 10^{-19}J. Recent measurement [59] on RBCs provides a bending rigidity of about 3 10^{-19}J.

- The capillary numbers associated with membrane plane-elasticity (shear and stretching)

$$C_s = \frac{\eta R\dot{\gamma}}{B} \equiv \dot{\gamma}\tau_s; \quad C_e = \frac{\eta R\dot{\gamma}}{C} \equiv \dot{\gamma}\tau_e \tag{21}$$

where τ_s and τ_e are time scales of relaxation of shearing and extensional/compresional modes.

- The degree of confinement

$$C_n = \frac{2R}{w} \tag{22}$$

- The viscosity contrast measuring the ratio between the viscosity of the internal fluid, η', to that of the external one η

$$\lambda = \frac{\eta'}{\eta} \tag{23}$$

- The reduced volume

$$\nu = \frac{V/(4\pi/3)}{[A/4\pi]^{3/2}} \tag{24}$$

which is the ratio of the actual area enclosed by the vesicles over the that of a circle having the same perimeter.

Using the values of parameters given above, taking for the typical radius of a human RBC $R \sim 3$ μm and using the plasma viscosity (close to that of water), we find for the capillary numbers (in c.g.s units): $C_\kappa \sim 0.1\dot{\gamma}$, $C_s \sim 10^{-3}$ $\dot{\gamma}$, $C_e \sim 10^{-5}\dot{\gamma}$. The small prefactor in C_e means that the relaxation time for extensional mode is of about 10^{-5} s., for shearing mode it is of about 10^{-3} s., and for bending mode 0.1 s. So the highest time scale seems to be the one associated with the bending mode, the most relevant one, in principle. This time scale is consistent with the study of Tomaiuolo and Guido in start-up shape dynamics of RBCs in microcapillary flow [60], where it is found that the time needed for a RBC to adapt a stationary shape after application of a flow is of the order of 0.1 s.

6. A vesicle, capsule and RBC under linear shear flow in an unbounded geometry

6.1. Vesicles

This is the simplest situation where the system is made of only a fluid membrane. Vesicles have revealed several types of motions under this type of flow, and the full phase diagram was given recently [7] by means of numerical simulations using the boundary integral formulation in three dimensions. Analytical [11, 12, 61] and semi-analytical [51] determinations of the phase diagram were made. If the shear flow is unbounded $C_n = 0$. Furthermore for a fluid membrane only the capillary number C_κ enters into play. For a fixed ν the parameter space is thus two-dimensional (C_κ and λ). At small enough C_κ (see figure 2) the motion is of tank-treading (TT) type, where the angle between the main axis of the vesicle and the flow direction remains constant in the course of time, while the membrane (which is fluid) undergoes a tank-treading motion. Upon increasing λ, the TT motion undergoes a saddle-node bifurcation towards tumbling (TB). If C_κ is not too small the TT motion becomes unstable via a Hopf bifurcation and an intermediate regime takes place, namely vacillating-breathing (VB) mode: the long axis does not make full rotation (unlike TB), but rather oscillates around the flow direction, while the shape undergoes breathing-like motion. The existence of this type of mode was predicted analytically [9]. The VB mode was called later, unfortunately, swinging [51] or trembling [11], and this has caused some semantic confusion in the literature. Upon increasing λ the VB mode becomes a TB mode; the angle of oscillation increases gradually until it reaches $\pi/2$ in total amplitude, where TB prevails. At higher λ, TB undergoes kayaking (a denomination already used before for ellipsoids, liquid crystals; the terminology "spinning" was also used in the case of vesicles [62]), where the main axis describes a cone about the perpendicular to the plane of the shear flow. For $\nu = 0.95$ the full numerical simulation has been captured by analytical theory [61] (using the quasi-spherical assumption, which is legitimate for small enough ν). By varying ν the phase diagram keeps the same kind of shape [7]. Experiments have been performed and have reported that there is a qualitative agreements with the theory and numerical simulation [63, 64]. However, experiments reported that the phase diagram could be described by only two dimensionless parameters, related to the above ones by

$$\Lambda = \frac{4\sqrt{\Delta}}{\sqrt{30\pi}}(1 + \frac{23\lambda}{32}), \quad S = \frac{7\pi}{3\sqrt{3}\Delta}C_a \tag{25}$$

with $\Delta = 4\pi(\nu^{-2/3} - 1)$, called the excess area with respect to a sphere (for a sphere $\nu = 1$ and thus $\Delta = 0$). Simulation as well as advanced analytical theory [61] showed that was not possible, three independent parameters seem necessary (C_a, λ, ν) (or Λ, S, Δ)). This has been confirmed by another simulation result [65].

6.2. Capsules and RBCs

Recent analytical calculations on inextensible capsules (using linear elasticity for the membrane, for the sake of simplicity) have been proposed, and they led to some interesting phenomena.

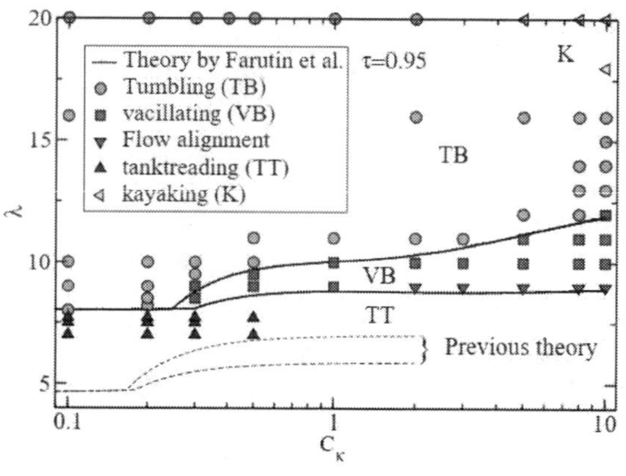

Figure 2. The phase diagram of a vesicle in a shear flow showing the various kinds of motions obtained numerically in three dimensions [7]. Shown is also a comparison with the analytical theory by Farutin et al.[13].

Besides the TT, VB and TB modes reported above, the shear elasticity introduces two main new effects: (i) if there is no viscosity contrast, under a weak enough flow (where a fluid vesicle would always exhibits TT), the capsule undergoes TB. The weak flow can not overcome the cytoskeleton pinning, since any motion of the network (considered to be relaxed in its reference state) will have to shear the network since the cytoskeleton has to accommodate, if it had to undergo TT, the cell geometry. Therefore the membrane undergos a TB motion, at the price of keeping the cytoskeleton close to its equilibrium configuration. Beyond a critical value of shear rate, the shear stress can overcome the resistance to network deformation, and the membrane shows a TT motion. (ii) In the TT motion, when a material point (an thus the cytoskeleton) moves, it remembers, so to speak, its original configuration after a half turn (for an ellipsoid, for example, after π rotation the shape is the same), and attempts to go back to that unstressed reference state. This attempt of relaxation (the associated time scale is fixed by the cytoskeleton dynamics) causes small oscillations of the angle between the main axis of the TT cell. These small oscillations were named swinging (SW), not to be confused with Vb where the oscillation is quite large accompanied with breathing. This kind of motion was reported experimentally in [66, 67]. If the viscosity contrast increases further and further, approaching the TB regime, the cell undergoes a motion which is very close to VB mode (ample amplitude breathing, while the main axis makes oscillation about the flow direction). A theoretical study was first undertaken by assuming that the cell has a fixed shape (following the seminal work of Keller and Skalak [68]) by adding the contribution due to the cytoskeleton [69] that has been successful in accounting for some of the motion of the RBC (especially SW). Later studies [16, 17] have relaxed the assumption of a fixed shape by following a previous study on vesicles [9]. The motions like SW was recovered, and the studies showed that the VB mode still manifests itself with ample amplitude. These studies have also pointed out that the so-called intermittent motion (SW interrupted by TB and so on) reported in Refs. [69, 67] was an artefact of the model (fixed shape), and that this motion disappears if the shape is free to evolve. This result [16, 17] is consistent with numerical simulations which showed no intermittency [70, 71, 72].

6.3. Impact on rheology

A natural question is how do the various types of dynamics described above affect rheology. This question was addressed first theoretically in the dilute regime for the vesicle system [9, 73, 10]. A first unexpected result is that the effective viscosity, in the TT regime, decreases upon increasing the viscosity contrast λ. More precisely, if we define the effective dimensionless viscosity as $\eta_{eff} = (\eta_s - \eta)/(\eta\phi)$, where η_s is the viscosity of the suspension, η that of the suspending fluid and ϕ the volume fraction, it was found in the quasi-spherical regime that [9]

$$\eta_{eff} = 1 + \frac{5}{2}\phi - \phi\frac{23\lambda + 32}{16\pi}\Delta \tag{26}$$

where recall that Δ is the excess area from a sphere (it vanishes for a sphere, in that a case a vesicle behaves like a rigid particle, and we recover the Einstein result). This result is valid for small enough ϕ (say few %). We see that η_{eff} decreases with λ, which is a completely opposite result to that of Taylor for emulsion

$$\eta_{eff} = 1 + \phi\frac{5\lambda/2 + 1}{\lambda + 1} \tag{27}$$

which increases with λ. A brief explanation is presented in [9], and a more extensive discussion is given in [74]. This results is valid in the TT regime only. Further increase of λ induces a transition from TT to VB or TB. In the TB regime the viscosity increases with λ [12]. In other words η_{eff} decreases in the TT regime and increases in the TB regime as λ is varied, meaning that η_{eff} exhibits a minimum at the TT-TB transition. This result was confirmed by numerical simulation in 2D [74] and in 3D [65]. It is interesting to note that the overall features of equation (26), which is derived in 3D, is also captured in 2D simulations. This points to the nontrivial fact that the 2D assumption is quite realistic in modeling rheology in the dilute regime. The same trend is found in the simulations of capsules [75]. Experiments have been performed on vesicles [76, 77] and RBCs suspensions [76]. The experiments in [76] show exactly the same trend. Other experiments in [77] arrived to the same conclusion as long as λ is of order of, or bigger, than 1. It was reported in that experiment that in the small λ regime, η_{eff} decreases with λ (as in the Taylor problem). This last result is not, so far, supported by numerical simulations nor by the experiments made in Ref.[76]. Another fact is that experiments on rheology of vesicle suspensions [76] have been performed for higher volume fractions (up to 12%) where hydrodynamic interactions can no longer be ignored. Nevertheless, the effective viscosity exhibits the same qualitative features as that obtained analytically [12] and numerically [74, 65]: it decreases with the viscosity contrast in the TT regime, goes through a minimum in the vicinity of the TT to TB bifurcation, and then increases in the TB regime (recall that expression .(26) is valid in the TT regime only).

7. Vesicles under a Poiseuille flow

7.1. Cross-streamline migration

The notion of lift force is an important field of research in mechanics. A spinning sphere is known to undergo a lift force due to inertia [78]. In the Stokes regime (relevant to blood flow), a spinning sphere does not feel any lift force due to the reversibility of the Stokes equation upon time reversal. However, a cell under shear flow is deformed and acquires an upstream/downstream asymmetry that destroys some symmetry (Stokes flow with boundary condition on a free surface, the cell one) under time reversal, that may gives rise to a lift force. As a consequence of thus in a Poiseuille flow (even in un unbounded geometry) a cell has the ability to exhibit cross-streamline migration. This mechanism is responsible (in part) for the Fahraeus-Linqvist effect (reported in 1931)(see [79]). Indeed, measurements in vitro show a quite counterintuitive effect: the apparent

Journal of Physics: Conference Series **392** (2012) 012005 doi:10.1088/1742-6596/392/1/012005

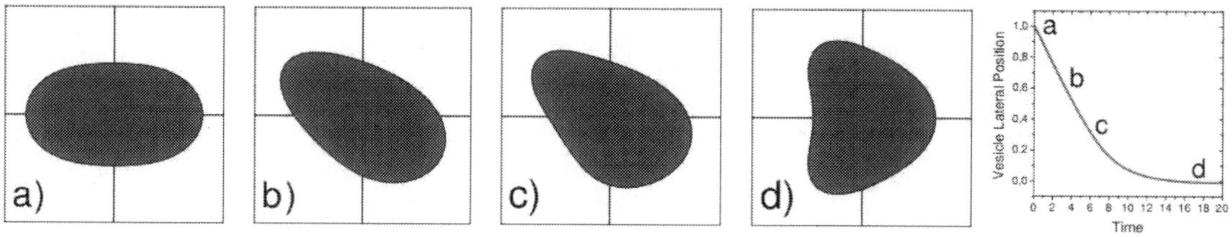

Figure 3. A vesicle which is initially off-centered may reach the center of the Poiseuille flow in the course of time. This happens if the flow strength is large enough. On the left side we see the various shapes exhibited by the vesicle as it moves towards the center. The final shape is a symmetric parachute.

blood viscosity decreases upon decreasing the channel diameter D, until a typical value of D of about few micrometers; in other words the more the blood is squeezed, the easiest it flows!. It is only when D becomes of the order of the RBC diameter that the apparent viscosity shows a sudden increase. The Fahraeus-Lindqvist effect is explained by the existence of a depletion layer (a layer which is free of RBCs) near the wall boundaries. The cell-free layer is due to (i) a lift force caused by the wall, and its origin is of purely viscous nature (unlike macroscopic lift -e.g. on a ball or airplane-which is due to inertia), (ii) a cross-streamline migration associated with the shear gradient of the Poiseuille flow. These two lift forces have been analyzed in the case of a single particle (namely a vesicle or capsule representing a model for the RBC) both theoretically [80, 20, 81, 6, 82, 10] and experimentally [83, 84, 85]. It was initially found numerically that a vesicles which is placed away from the centerline of the Poiseuille flow migrates towards the center, as shown in figure 3. It was suggested [6] that the migration velocity scales as

$$v_{migration} \sim \alpha f(C_\kappa(y)) \tag{28}$$

where α is the curvature of the Poiseuille flow and f is a universal function of the capillary number, where $C_\kappa(y)$ is the local capillary number in which $\dot{\gamma}$ is replaced by the local gradient of the Poiseuille flow. Using the small deformation theory for vesicles [86] it was found that far away from the center the migration velocity is constant and is given by

$$v_{migration} \sim \alpha \Delta^{1/2} R^2 \tag{29}$$

and does not depend on membrane rigidity. The prefactor is a complicated function of the viscosity ratio λ. Experiments on vesicles have analyzed the migration in a microfluidic channel and it was shown that the vesicle always migrates towards the center [85]. However, in a very weak flow (much weaker than usual flows in experiments), it has been shown that a quasi-spherical vesicle can migrate outwards [87].

7.2. Symmetry-breaking: slipper

A longstanding puzzle in blood microcirculation is the understanding of why do RBCs adopt an asymmetric shape (called a slipper) in microcirculation, despite a symmetric flow. In vivo the volume fraction of RBCs in the microvasculature (also called hematocrit) is of about 20% (unlike in macrocirculation where the hematocrit is of about 45%), so that RBCs are in a close contact with each other. The first comprehensive experimental report on the slipper phenomenon is that of Skalak and Branemark[88] (who studied a cluster of cells), followed by another experimental study by Gaehtgens et al. [89]. Subsequent studies on single RBCs exhibiting a slipper shape in a glass tube were briefly reported by Secomb et al. [90], and analyzed in more detail by Guido and Tomaiuolo [91]. This problem was revisited more recently [92, 93].

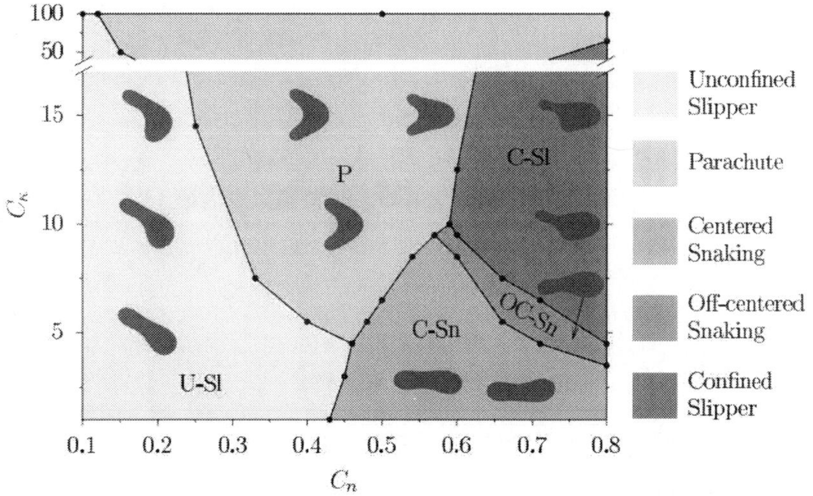

Figure 4. The phase diagram of a vesicle in a Poiseuille flow showing 5 distinct regions.

A first theory, using a lubrication approximation was proposed by Secomb and co-workers [94], and the slipper shape was later observed in numerical simulations by Pozrikidis [95]. Two dimensional simulations were considered more recently by Secomb et al. [90], who reported on the slipper shape. Recently, considering [96] an unbounded parabolic velocity profile, the slipper shape was shown to be an intrinsic configuration resulting from the instability of the symmetric (parachute) shape.

More recently [97] the study of the slipper in the presence of bounding walls has been considered. The walls did not destroy the branch of solution found in the unbounded geometry [96]. However the presence of walls has led to an unexpected complexity. Five distinct morphologies and dynamics are identified, depending on flow and structural parameters. The viscosity contrast has been set to unity and the reduced area $\nu = 0.6$ (close to that of the Human RBC), and the two other parameters C_κ and C_n have been varied. The shape and dynamics phase diagram is presented in figure 4 where there are 5 distinct regions with different types of shapes. For a weak and strong confinement, there is always a slipper solution, the intermediate confinement shows either a symmetric parachute shape, or a motion of flagella type, that has been called "snaking". Figure 5 shows a snapshot of the snaking motion (figure 5). The snaking may be either centered (the center of mass undergoes oscillation with a mean value equal to zero), or non-centered (the mean value of the center of mass is different from zero).

The phase diagram shown in Fig. 4 says that for a given confinement C_n, the transition from the parachute shape to a slipper one always occurs upon *decreasing* the flow strength (which corresponds to decreasing C_κ). Experiments are sparse [92, 91, 93], but there is a consensus that the occurrence of the slipper occurs upon *increasing the flow strength*, which is in contradiction with the phase diagram shown in Fig. 4. We have attempted to understand this major discrepancy, and came to the conclusion that the viscosity contrast plays a major role. RBCs in vivo have a viscosity contrast of about $\lambda \simeq 5$, and this ratio depends on temperature. It has been found very recently [98] that introduction of a viscosity contrast leads to new branches of slipper solutions that corresponds to a transition from parachute to slipper upon increasing the flow strength, in accord with the experiments.

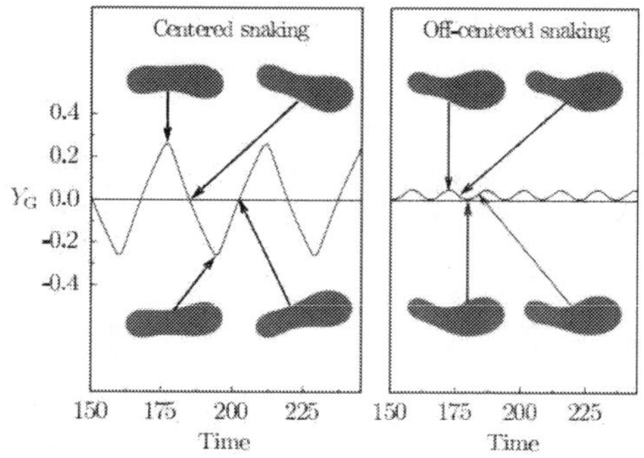

Figure 5. A snapshot of the snaking motion, where the lines show the position of the center of mass as a function of time.

8. Some collective motions in a Poiseuille flow

Experiments on RBCs suspensions are known to form rouleaux at low enough shear rate, and are destructed upon increasing flow strength. The rouleaux formation is mediated by the so-called depletion forces (osmosis-type of attraction) caused by plasma proteins (fibrinogen and β-globuline)[99]. Actually, numerical simulations of RBCs in a channel flow (in the absence of any interaction mimicking the depletion force) form clusters of purely hydrodynamical origin [100, 101]. The cluster formation has been analyzed in details experimentally [102]. The analysis of the flow field around a cell show that the Poiseuille flow is modified by the cell which creates converging vortices at the front (aspiring thus a cell ahead of it; see figure 6) and diverging at the rear, avoiding thus cells to come too close to each other. The size of the cluster is a function of the flow rate [101]. It is shown that for a given flow rate there exists a maximum cluster size, so that if the initial formed cluster is too large, it will expel one or few cells ahead in order to bring down the size of the cluster to the marginally stable one. Many other other collective phenomena are currently studied and will give rise to publications in the next few months.

9. Conclusion

This review was intended to present a brief description of some aspects of the field of vesicles, capsules and RBC dynamics under flow, and to show the impressive upsurge of interest in the past few years. This short review does not pretend to have covered all phenomena studied recently in this field, but at least it provides a large scale picture of the recently realized and ongoing research in this field. We are at the prelude of being able to simulate with realistic models a large number of cells (with blood sample of few centiliters) thanks to the progress of computational science (parallel, GPU –Graphic Processing Units – domain decomposition theories, highly performing finite element libraries...). My personal guess is that it will be possible to simulate a real blood flow in a real architecture in animals (like mouse) within the next ten to twenty years, which will then open the way to simulation of the circulatory system of the human body. This field will surely grow rapidly further and further and the next research generation will be mainly directed towards implication in health problems and care: physical factors impacting cardiovascular diseases and relation to blood flow, pressure waves and relation to blood tension, capillary rarefaction, thrombosis, aneurysms, the role of endothelium, coupling to arterial mechanics, the role of glycocalyx (a polymer brush covering the interior of blood vessels), the lab-on-chip technologies (blood diagnosis, blood sorting out...), and so on.

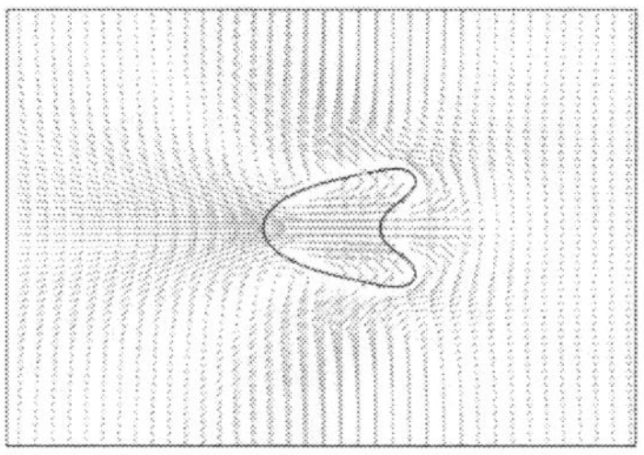

Figure 6. The induced flow field pattern in the comoving frame showing converging vorticies ahead and diverging ones at the rear. This explains the origin of clustering.

9.1. Acknowledgments
The author would like to acknowledge financial support from CNES (Centre National d'Etudes Spatiales) and ESA (European Space Agency) and ANR (Agence National de la Recherche), project "MOSICOB".

[1] Helfrich W 1973 *Z. Naturforschung* **28** 693–703
[2] Seifert U 1997 *Advances in Physics* **46** 13–137
[3] Barthès-Biesel D 2009 *C.R. Physique* **10** 764–774
[4] Laadhari A, Saramito P and Misbah C 2012 *Phys. Fluid* **24** 031901–031907
[5] Laadhari A, Misbah C and Saramito P 2010 *Physica D* **239** 1567–1572,
[6] Kaoui B, Ristow G H, Cantat I, Misbah C and Zimmermann W 2008 *Phys. Rev. E* **77** 021903
[7] Biben T, Farutin A and Misbah C 2011 *Phys. Rev. E* **83**(3) 031921
[8] Skalak R, Tozeren A, Zarda R and Chien S 1973 *Biophys. J.* **13** 245–264
[9] Misbah Ç 2006 *Phys. Rev. Lett.* **96** 028104
[10] Vlahovska P M and Gracia R S 2007 *Phys. Rev. E* **75** 016313 (pages 12)
[11] Lebedev V V, Turitsyn K S and Vergeles S S 2007 *Phys. Rev. Lett.* **99** 218101
[12] Danker G, Biben T, Podgorski T, Verdier C and Misbah C 2007 *Phys. Rev. E* **76** 041905
[13] Farutin A and Misbah C 2011 *Phys. Rev. E* **84**(1) 011902
[14] Barthès-Biesel D and Rallison J M 1981 *J. Fluid Mech.* **113** 251–267
[15] Vergeles S and Vorobev P 2011 *JETP Lett.* **94** 513–518
[16] Vlahovska P, Young Y, Danker G and Misbah C 2011 *J. Fluid Mech.* **678** 221–247
[17] Finken R, Kessler S and Seifert U 2011 *Phys. Cond. Mat.* **23** 184113
[18] Gires P, Danker G and Misbah C 2012 *Phys. Rev. E (accepted)*
[19] Pozrikidis C 1992 *Boundary Integral and Singularity Methods for Linearized Viscous Flow* (Cambridge University Press, Cambridge, UK)
[20] Cantat I and Misbah C 1999 *Phys. Rev. Lett.* **83** 235–238
[21] Veerapaneni S K, Gueyffier D, Zorin D and Biros G 2009 *J. Comp. Phys.* **228** 2334–2353
[22] Kraus M, Wintz W, Seifert U and Lipowsky R 1996 *Phys. Rev. Lett.* **77** 3685–3688
[23] Lac E, Morel A and Barthès-Biesel D 2007 *J. Fluid Mech.* **573** 149–169
[24] Ying L, Biros G and Zorin D 2004 *J. Comp. Phys.* **196** 591–626
[25] Wang H, Lei T, Huang L and Yao Z 2004 *Int. J. Numer. Methods Eng.* **70** 812–839
[26] Ghigliotti G, Selmi H, Kaoui B, Biros G and Misbah C 2009 *Esaim: proceedings* **28** 212–227
[27] Peskin C 1977 *J. Comp. Phys* **25** 220–252
[28] Miller L and Peskin C 2005 *J. Exp. Biol.* **208** 195–212
[29] Doddi S K and Bagchi P 2009 *Phys. Rev. E* **79**(4) 046318
[30] Yazdani A Z K and Bagchi P 2011 *Phys. Rev. E* **84**(2) 026314
[31] Wang X, Zhang L and Liu W K 2009 *J. Comp. Phys* **228** 2535–2551
[32] Tsubota K i and Wada S 2010 *Phys. Rev. E* **81**(1) 011910
[33] Ismail M and Lefebvre-Lepot A 2012 *arXiv:1202.3034v1 [math-ph]*
[34] Biben T and Misbah C 2003 *Phys. Rev. E* **67** 031908
[35] Biben T, Kassner K and Misbah C 2005 *Phys. Rev. E* **72** 041921
[36] Du Q, Liu C and Wang X 2004 *J. Comp. Phys.* **198** 450–468
[37] Du Q, Liu C and Wang X 2006 *J. Comp. Phys.* **212** 757–777
[38] Jamet D and Misbah C 2008 *Phys. Rev. E* **78**(4) 041903

[39] Salac D and Miksis M 2011 *J. Comp. Phys.* **230** 8192–8215
[40] Maitre E, Misbah C, Peyla P and Raoult A 2012 *Physica D* **243** 1146–1157
[41] Hoogerbrugge P and Koelman J 1992 *Europhys. Lett.* **19** 155–160
[42] Espagnol P and Warren P 1992 *Europhys. Lett.* **30** 191–196
[43] Pivkin I V and Karniadakis G E 2008 *Phys. Rev. Lett.* **101**(11) 118105
[44] Fedosov D A, Pan W, Caswell B, Gompper G and niadakis G E K 2011 *PNAS* **108** 11772 11777
[45] Malevanets A and Kapral R 1999 *J. Chem. Phys.* **110** 8605–8613
[46] Ihle T and Kroll D M 2003 *Phys. Rev. E* **67**(6) 066705
[47] Kikuchi N, Pooley C, Ryder J and Yeomans J 2003 *J. Chem. Phys.* **119** 6388–6395
[48] Lamura A, Gompper G, Ihle T and roll D K 2001 *Europhys. Lett.* **56** 319–325
[49] Allahyarov E and Gompper G 2002 *Phys. Rev. E* **66**(3) 036702
[50] Ripoll M, Mussawisade K, Winkler R and Gompper G 2003 *Eurohys. Lett.* **68** 6388–6395
[51] Noguchi H and Gompper G 2007 *Phys. Rev. Lett.* **98**(12) 128103
[52] Noguchi H and Gompper G 2005 *PNAS* **102** 14159–14164
[53] McWhirter J L, Noguchi H and Gompper G 2009 *PNAS* **106** 6039–6043
[54] Dupin M M, Halliday I, Care C M, Alboul L and Munn L L 2007 *Phys. Rev. E* **75**(6) 066707
[55] Zhang J, Johnson P C and A S Popel P B 2007 *Phys. Biol.* **4** 285–285
[56] Kaoui B, Harting J and Misbah C 2011 *Phys. Rev. E* **83**(6) 066319
[57] Clausen J and Aidun C 2010 *Phys. Fluid* **23** 123302–123311
[58] Ghigliotti G, Rahimian A, Biros G and Misbah C 2011 *Phys. Rev. Lett.* **106**(2) 028101
[59] Betz T, Lenz M, Joanny J F and Sykes C 2009 *PNAS* **106** 15320–15325
[60] Guido S and Tomaiuolo G 2009 *Microvascular Research* **82** 35–41
[61] Farutin A, Biben T and Misbah C 2010 *Phys. Rev. E* **81**(6) 061904
[62] Lebedev V, Turitsyn K and Vergeles S 2008 *New J. Phys.* **10** 043044 (35pp)
[63] Deschamps J, Kantsler V, Segre E and Steinberg V 2009 *PNAS* **106** 11444–11447
[64] Deschamps J, Kantsler V and Steinberg V 2009 *Phys. Rev. Lett.* **102** 118105 (pages 4)
[65] Zhao H and Shaqfeh E 2011 *J. Fluid Mech.* **674** 578–604
[66] Bitbol M 1986 *Biophys. J.* **49** 1055–1068
[67] Abkarian M, Faivre M and Viallat A 2007 *Phys. Rev. Lett.* **98**(18) 188302
[68] Keller S and Skalak R 1982 *J. Fluid Mech.* **120** 27–47
[69] Skotheim J M and Secomb T W 2007 *Phys. Rev. Lett.* **98**(7) 078301
[70] Kessler S, Finken R and Seifert U 2008 *J. Fluid Mech.* **605** 20
[71] Sui Y, Chew Y, Roy P, Cheng Y and Low H 2008 *J. Fluid. Mech.* **20** 112106
[72] Bagchi P and Murthy K R 2009 *Phys. Rev. E* **80**(1) 016307
[73] Danker G and Misbah C 2007 *Phys. Rev. Lett.* **98** 088104
[74] Ghigliotti G, Biben T and Misbah C 2009 *J. Fluid Mech.* **653** 489–518
[75] Bagchi P and Murthy K R 2010 *Phys. Rev. E* **81**(5) 056320
[76] Vitkova V, Mader M A, Polack B, Misbah C and Podgorski T 2008 *Biophys. J.* **95** 33–35
[77] Kantsler V, Segre E and Steinberg V 2008 *Europhys. Lett.* **82** 58005
[78] Feuillebois F 1989 *Multiphase Sci. Technol.* **4** 583
[79] Pries A, Neuhaus N and Gaehtgens P 1992 *Am. J. Physiol.* **20** H1770–H1778
[80] Olla P 1997 *J. Phys. A: Math. Gen.* **30** 317–329
[81] Sukumaran S and Seifert U 2001 *Phys. Rev. E* **64**(1) 011916
[82] Doddi S K and Bagchi P 2008 *Int. J. Multiphase Flow* **34** 966–986
[83] Abkarian M, Lartigue C and Viallat A 2002 *Phys. Rev. Lett.* **88** 068103
[84] Callens N, Minetti C, Coupier G, Mader M, Dubois F, Misbah C and Podgorski T 2008 *Europhys. Lett.* **83** 24002
[85] Coupier G, Kaoui B, Podgorski T and Misbah C 2008 *Phys. Fluids* **20** 111702
[86] Danker G, Vlahovska P M and Misbah C 2009 *Phys. Rev. Lett.* **102** 148102 (pages 4)
[87] Farutin A, Biben T and Misbah C 2010 *Phys. Rev. E* **81**(6) 061904
[88] Skalak R and Branemark P I 1969 *Science* **164** 717–719
[89] P Gaehtgens C D and Albrecht K 1980 *Blood Cells* **6** 799–812
[90] Secomb T, Styp-Rekowska B and Pries A 2007 *Ann. Biomed. Eng.* **35** 755–765
[91] Tomaiuolo G, Simeone M, Martinelli V, Rotoli B and Guido S 2009 *Soft Matter* **5** 3736–3740
[92] Abkarian M, Faivre M, Horton R, Smistrup K, Best-Popescu C A and Stone H A 2008 *Biomed. Mater.* **3**(7) 034011–034025
[93] Guido S and Tomaiuolo G 2009 *C.R. Physique* **10** 752–763
[94] Secomb T W and Skalak R 1982 *Microvasc. Res.* **24** 194–203
[95] Pozrikidis C 2005 *Ann. Biomed. Eng.* **33** 165–178

[96] Kaoui B, Biros G and Misbah C 2009 *Phys. Rev. Lett.* **103** 188101
[97] Kaoui B, Tahiri N, Biben T, Ez-Zahraouy H, Benyoussef A, Biros G and Misbah C 2011 *Phys. Rev. E* **84**(4) 041906
[98] Ez-Zahraouy N T H, Biben T, Benyoussef A and Misbah C 2012 On the problem of slipper shapes of red blood cells in the microvasculaure
[99] Thurston G B 2006 *Biorhelogy* **43** 729
[100] McWhirter J L, Noguchi H and Gompper G 2011 *Soft Matter* **7**(22) 10967–10977
[101] Ghigliotti G, Selmi H, Elasmi L and Misbah C 2012 *Phys. Fluid. (submitted)*
[102] Tomaiuolo G, Lanotte L, Ghigliotti G, Misbah C and Guido S 2012 *Phys. Fluid* **5** 051903

Computation of Dynamics of Soft Particles using Front-Tracking Method

Masato Makino and Osamu Sano

Department of Applied Physics, Tokyo University of Agriculture and Technology, Koganei, Tokyo 184-8588, Japan
JST-CREST, Chiyoda, Tokyo 102-0075, Japan

E-mail: `sano@cc.tuat.ac.jp`

Abstract. The dynamics of droplets and elastic capsules in a low Reynolds number flow is investigated by Front-Tracking method. The motions of droplets under gravity field, shear flow and Poiseuille flow are shown as examples. The capsules are migrated to the center line of the Poiseuille flow due to the lateral migration. The hydrodynamic conductance of the Poiseuille flow is reduced as the capsule goes to the center of the channel. The small capsules move to the boundary wall of the Poiseuille flow in the mixture of large and small capsules.

1. Introduction

The motion of deformable particles, such as droplets, vesicles and capsules, is becoming one of the most remarkable study. The simulation of the deformable particles are well presented in various literatures[1, 2, 3]. The study of the motion of deformable particles provides us useful information for the development of a Drug Delivery System (DDS). The suction of a droplet from simple shear flow region to Poiseuille flow region has been dealt with in our previous paper [4], as shown in Fig. 1. We have elucidated the criterion of the suction for the enhancement of the efficiency of drug transportation to the diseased part.

In general, the DDS particle flows in the blood vessel dispersed with red blood cells, white blood cells, platelets, and so on. The DDS particle we consider is smaller than the red blood cell. The blood cell has a tendency to move to the center of the vessel because of the lateral migration. As a consequence, the DDS particle is anticipated to be pushed out from the central region toward the periphery of the blood vessel. Then the DDS particle will be more easily sucked into the cleft of the vessel to cure the inflamed part.

In this paper, a simulation of the deformable particles under gravity or pressure gradient field is shown. In the simulation, the finite difference method and Front-Tracking method are used. The method was developed by Peskin *et al.*[5, 6, 7, 8] and was applied for droplet migration problems by Tryggvason *et al.*[9, 10, 11, 12, 13, 14, 15, 16, 17]. The Front-Tracking method uses two types of mesh, Eulerian mesh and Lagrangian mesh. The Eulerian mesh is used to calculate the field data of velocity and pressure. The Lagrangian mesh describes the shape of deformable particles. Examples of the droplets under gravity, the droplets under shear field, and the elastic capsules in the Poiseuille flow are presented in this study.

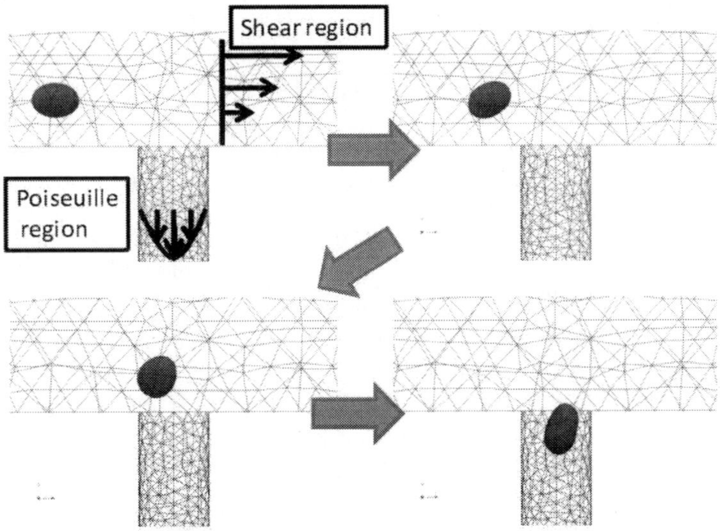

Figure 1. Suction of a droplet from the shear region to the Poiseuille region.

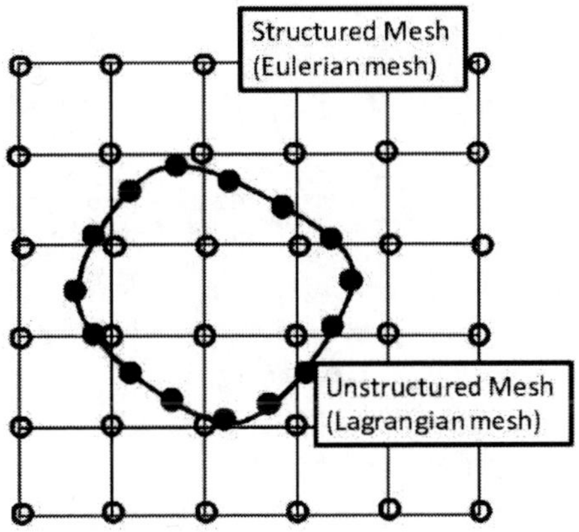

Figure 2. Lagrangian mesh, which composes a particle shape, is immersed in Eulerian mesh.

2. Formulation

In this section, we describe the method of the computation of deformable particles in the Newtonian fluid.

A set of rectangular structured mesh (Eulerian mesh) for fluid computation and unstructured triangular mesh (Lagrangian mesh) for a deformable particle is prepared as shown in Fig. 2. In general, a density field and a viscosity field depend on whether the Eulerian mesh points are inside of the particle or outside of the particle.

The velocity field $\boldsymbol{v}(\boldsymbol{r})$ and pressure field $p(\boldsymbol{r})$ of an incompressible fluid with density $\rho(\boldsymbol{r})$ and viscosity $\mu(\boldsymbol{r})$ are governed by the Navier-Stokes equation

$$\rho\left(\frac{\partial \boldsymbol{v}}{\partial t} + \boldsymbol{v}\cdot\nabla\boldsymbol{v}\right) = -\nabla p + \nabla\cdot\mu(\boldsymbol{r})\left(\nabla\boldsymbol{v} + (\nabla\boldsymbol{v})^t\right) + \rho\boldsymbol{g} + \boldsymbol{f}, \tag{1}$$

and the continuity equation

$$\nabla\cdot\boldsymbol{v} = 0, \tag{2}$$

where $\boldsymbol{A}^t$ denotes the transpose of tensor $\boldsymbol{A}$, $\boldsymbol{g}$ is a gravity acceleration and $\boldsymbol{f}$ is the force acting from the interface of deformed particle. The fluid outside of the droplet has a viscosity $\mu(\boldsymbol{r}) = \mu$ and a density $\rho(\boldsymbol{r}) = \rho$, whereas the fluid inside the droplet has a viscosity $\mu(\boldsymbol{r}) = \mu'$ and a density $\rho(\boldsymbol{r}) = \rho + \Delta\rho$. The viscosity ratio λ is defined as $\lambda = \mu'/\mu$.

The velocity and pressure are assigned on the Eulerian mesh point. The Highly Simplified Maker And Cell(HSMAC) method is used to solve the fluid fields. The time evolution of the front of the particles is calculated from the velocity on the surface. Because the mesh point (Lagrangian mesh) on the surface does not coincide with the field point(Eulerian mesh), the velocity $\boldsymbol{v}^L$ on the surface is calculated from the field velocity $\boldsymbol{v}^E$ as

$$\boldsymbol{v}^L(\boldsymbol{r}) = \sum w(\boldsymbol{r} - \boldsymbol{r}')\boldsymbol{v}^E(\boldsymbol{r}') \tag{3}$$

where $w(\boldsymbol{r})$ is defined as

$$w(\boldsymbol{r} - \boldsymbol{r}') = d\left(\frac{r_x - r'_x}{\Delta x}\right) d\left(\frac{r_y - r'_y}{\Delta y}\right) d\left(\frac{r_z - r'_z}{\Delta z}\right), \tag{4}$$

$\Delta x, \Delta y$ and Δz are the sizes of the lattice for each direction, and Brackbill and Ruppel's weight function [18]

$$d(r) = \begin{cases} \frac{2}{3} - r^2 + \frac{1}{2}r^3, & 0 < r < 1 \\ \frac{1}{6}(2 - r)^3, & 1 \leq r \leq 2 \\ 0, & otherwise \end{cases} \tag{5}$$

is used in our simulation.

On the other hand, the force density $\boldsymbol{f}$ in eq. (1) is calculated by

$$\boldsymbol{f}(\boldsymbol{r}') = \sum w(\boldsymbol{r}' - \boldsymbol{r})\boldsymbol{t}(\boldsymbol{r})\frac{\Delta S}{\Delta x \Delta y \Delta z} \tag{6}$$

where $\boldsymbol{t}$ is a traction on the surface of the particle and ΔS is the area of the surface element.

The traction $\boldsymbol{t}$ depends on the expression of the surface free energy. For a droplet with the surface tension coefficient σ, the surface free energy F^d is

$$F^d = \sigma \int dS, \tag{7}$$

where the integration is over the droplet surface. The traction is derived from the functional derivation

$$\boldsymbol{t} = \frac{\delta F^d}{\delta \boldsymbol{r}} = 2\sigma H \boldsymbol{n}, \tag{8}$$

where H is the mean curvature and $\boldsymbol{n}$ is a unit normal on the droplet.

For the elastic membrane with Young's modulus E, the free energy F^e is

$$F^e = \frac{E}{6} \int \left(\lambda_1^2 + \lambda_2^2 + \lambda_1^{-2}\lambda_2^{-2} - 3\right) dS, \tag{9}$$

where λ_1 and λ_2 are the square roots of eigen values of Cauchy-Green tensor. For the details, consult literatures [3] and [19].

The position $\boldsymbol{r}(t + \Delta t)$ of the node on the particle surface is evolved by the Euler method

$$\boldsymbol{r}(t + \Delta t) = \boldsymbol{r}(t) + \boldsymbol{v}(t)\Delta t. \tag{10}$$

For the droplet, the mesh stabilization algorithm, which is described in a paper [20], is used. Figure 3 shows the efficiency of this algorithm. The connection of the nodes does not change, and the calculation is stable as shown in Fig. 3 (b).

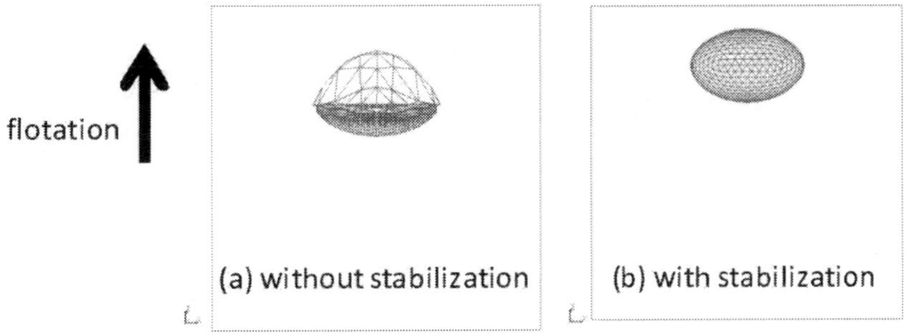

Figure 3. Mesh stabilization for a droplet. Fig. 3(a) shows a simulation without the mesh stabilization. The nodes on the mesh flows toward the rear side of the droplet. Fig. 3(b) shows a simulation with the mesh stabilization.

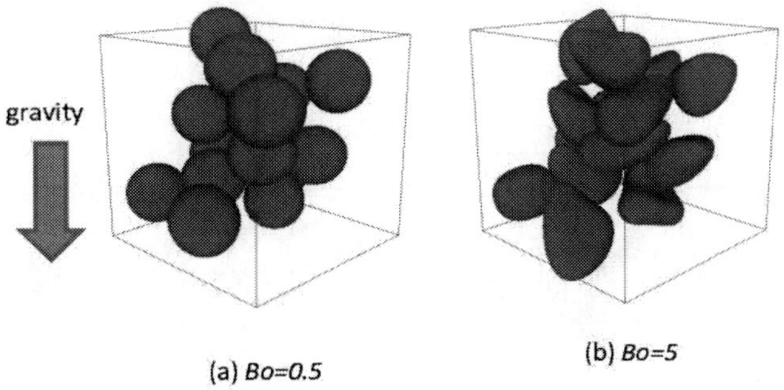

Figure 4. Sedimentation of droplets. Bond number Bo is 0.5 in (a) and 5 in (b).

3. Results

3.1. Sedimentation of droplets

The snapshots of the droplets under gravity with $Re = 23$ are shown in Fig. 4. The periodic boundary is used in this simulation. Bond number is defined as the ratio of the gravity to the surface tension $Bo = \Delta\rho g a^2/\sigma$, where $\Delta\rho$ is the density difference between the bulk fluid and the droplet and a is the radius of the droplet. When the Bond number is higher, the gravity effect is higher than the surface tension so that the droplet is more deformable. The droplets in Fig. 4(a) keep spherical shapes, whereas the droplets in Fig. 4(b) are deformed under gravity. The average position of the droplets against time is shown in Fig. 5. The time evolution of the average position for $Bo = 0.5$ and 5 are almost the same. The movement of the droplets is not remarkably dependent on the deformation in this case. The inserted triangle in Fig. 5 shows the analytical reduced velocity in the case of a zero Reynolds number flow without interaction and deformation

$$\frac{V}{a^2\Delta\rho g\mu} = \frac{2(1+\lambda)}{3(2+3\lambda)}, \tag{11}$$

which is known as Hadamard-Rybczinski result. Our results show the multi-droplets motion is slightly faster than the single ideal motion of a droplet.

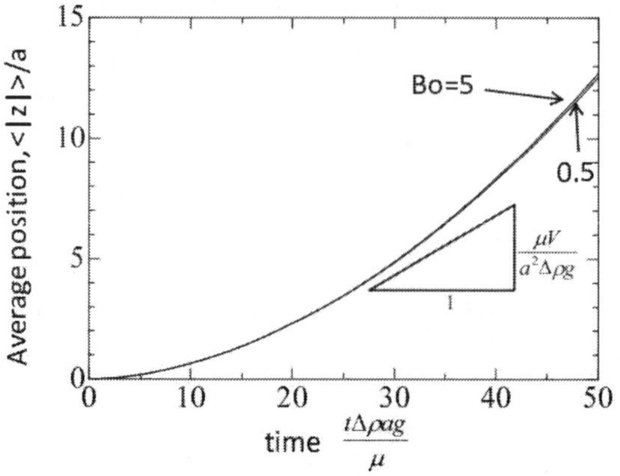

Figure 5. Average of the centers $< |z| > /a$ of the droplet as a function of time $t\Delta\rho a g/\mu$. A slope of inserted triangle in the figure shows the reduced velocity for a single spherical droplet at $Re = 0$.

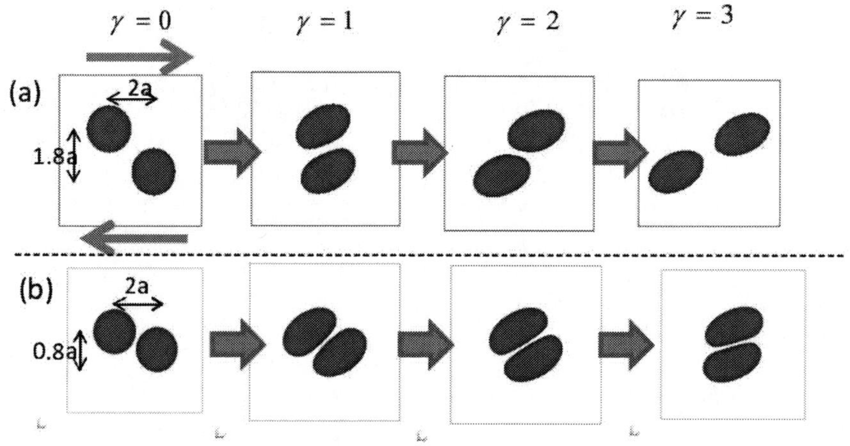

Figure 6. Shear induced motion of droplets. The distance between the centers of the droplets in the y direction is $1.8a$ in (a) and $0.8a$ in (b).

3.2. Shear induced deformation of droplets

Two droplets under shear flow with shear rate $\dot{\gamma}$ are shown in Fig. 6. In this case, Capillary number is $Ca = \mu\dot{\gamma}a/\sigma = 0.1$ and the viscosity ratio λ is 10. The initial distance between the centers of the droplets in the y direction is $1.8a$ in (a) and $0.8a$ in (b). The deformation of the droplets in (b) is larger than those in (a).

3.3. Lateral migration

An elastic capsule, whose property is given in eq.(9) with $a^2E/p_x = 1.28 \times 10^4$, is placed in the Poiseuille flow between the two plates with the separation distance $l_y = 3.2a$, where a is the radius of the capsule. The pressure gradient is p_x in the x direction and the Reynolds number Re is 0.1. The periodic boundary condition is imposed on the x and z directions. We assume

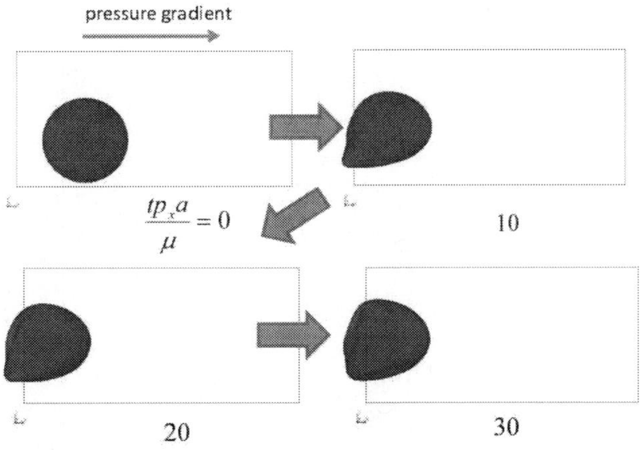

Figure 7. Lateral migration of an elastic capsule.

the analytic solution of stable Poiseuille flow as an initial velocity profile. Figure 7 shows the snapshots of the elastic capsule whose initial position from the bottom plate is $1.1a$. The capsule moves to the center of the channel. This phenomenon is known as the lateral migration.

Figure 8 shows the time variation of the center of the capsule with different initial positions. Every capsule moves to the center of the channel.

The hydrodynamic conductance is shown in Fig. 9. The fluid flux q is calculated by the integral of the x component's velocity v_x over the cross section S including inside the capsule, defined as

$$q = \frac{1}{S} \int_S v_x dS. \tag{12}$$

The hydrodynamic conductance is defined as $\nu = q/p_x$. The conductance ν_0 without capsules is $\nu_0 = l_y^2/(12\mu)$. Each line in Fig.9 is calculated for different initial position of the capsule. The conductance is low in the case where the capsule is near the wall, but it is high when the capsule is located in the central region. Due to the lateral migration, the terminal conductance is independent of the initial position of the capsule.

Finally, the oblate ellipsoidal capsules and the small spherical capsules are placed with the initial configuration as shown in Fig. 10(a). After the pressure gradient is applied, the large capsules remain in the center of the channel, whereas the small capsules move toward the boundary wall (See Fig. 10(b)). In general, the DDS particle is assumed smaller than the red blood cells. Then, the DDS particle moves toward the wall of the blood vessel, which makes it easy to reach the inflammation or a diseased part through the cleft of the blood vessel.

4. Summary

A simulation of the Front-Tracking method was presented in this paper. The motion of the droplet under the gravity field and the shear flow were shown. Lateral migration of elastic capsules was examined. The elastic capsules of the same size moved to the center of the Poiseuille flow. In the mixture of the small and large capsules, the large capsules moved to the center, whereas the small capsules moved toward the boundary walls. In this study, we showed the simulation method to obtain the fundamental information to design the DDS particle and biological cells.

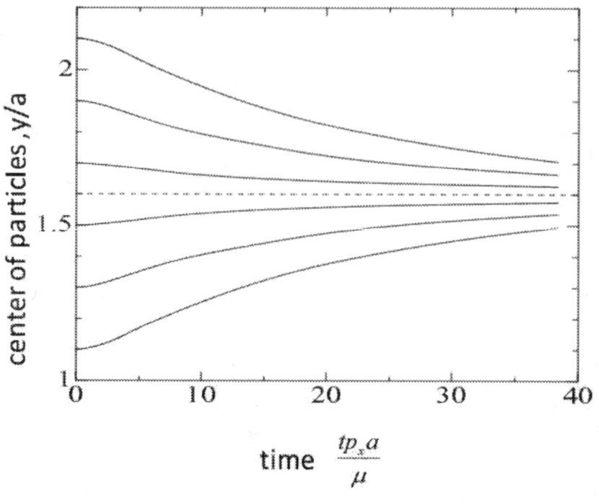

Figure 8. Center of the capsule y/a starting from different initial positions as a function of time tp_xa/μ. The dashed line indicates the center of the channel.

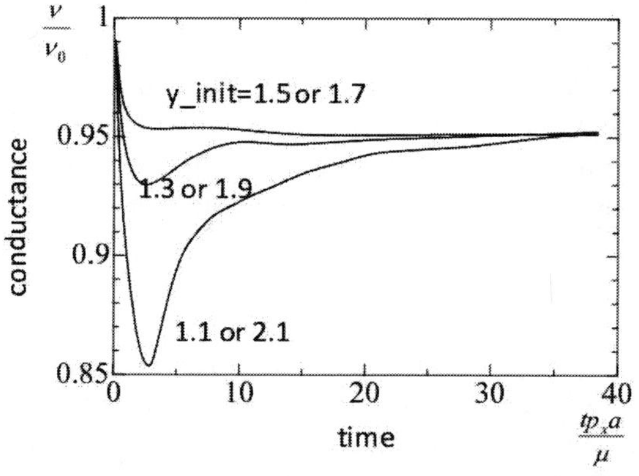

Figure 9. Hydrodynamic conductance ν/ν_0 of the channel as a function of tp_xa/μ.

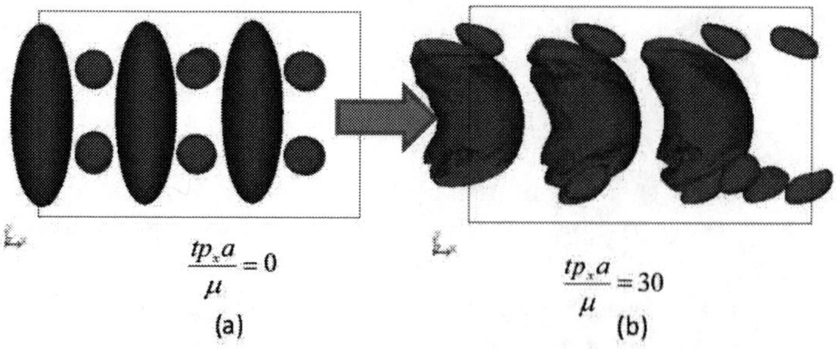

Figure 10. Small capsules move toward the wall of the channel, whereas large ones remain near the center of the channel.

Acknowledgments
This work is supported by the Japan Science and Technology Agency (JST/CREST). The authors used the GOURMET in the OCTA (http://octa.jp) for the data management and the visualization.

References

[1] Barthès-Biesel D, Yamaguchi T, Ishikawa T and Lac E 2006 From passive motion of capsules to active motion of cells *J. Biomech. Sci. and Engin.*
[2] Pozrikidis C 2003 *Modeling and Simulation of Capsules and Biological Cells* (New York:Chapman & Hall/CRC)
[3] Pozrikidis C 2010 *Computational Hydrodynamics of Capsules and Bilogical Cells* (New York:Chapman & Hall/CRC)
[4] Makino M and Sano O Motion of droplets under shear flow with suction *Fluid Dyn. Res.* (accepted for publication)
[5] Peskin C S 1977 Numerical analysis of blood flow in the heart *J. Comput. Phys.* **25** 220
[6] Peskin C S and McQueen D.M 1989 A three-dimensional computational method for blood flow in the heart I. Immersed elastic fibers in a viscous in compressible fluid *J. Comput. Phys.* **81** 372
[7] Peskin C S and Printz B.F 1993 Improved volume conservation in the computation of flows with immersed boundaries *J. Comput. Phys.* **105** 33
[8] Peskin C S 2002 The immersed boundary method *Acta Numer.* **11** 479
[9] Tryggvason G, Scardovell R and Zaleski S 2011 *Direct numerical simulations of gas-liquid multiphase flows* (New York:Cambridge University Press)
[10] Bunner B and Tryggavason G 2002 Dynamics of homogeneous bubbly flows Part 1. Rise velocity and microstructure of the bubbles *J. Fluid. Mech.* **466** 17
[11] Bunner B and Tryggavason G 2002 Dynamics of homogeneous bubbly flows Part 2. Velocity fluctuations *J. Fluid. Mech.* **466** 53
[12] Bunner B and Tryggavason G 2003 Effect of bubble deformation on the properties of bubbly flows *J. Fluid. Mech.* **495** 77
[13] Esmaeeli A and Tryggavason G 1998 Direct numerical simulations of bubbly flows. Part 1. Low Reynolds number arrays *J. Fluid. Mech.* **377** 313
[14] Esmaeeli A and Tryggavason G 1999 Direct numerical simulations of bubbly flows. Part 2. Moderate Reynolds number arrays *J. Fluid. Mech.* **385** 325
[15] Esmaeeli A and Tryggavason G 2005 A direct numerical simulation study of the buoyant rise of bubbles at O(100) Reynolds number *Phys. Fluids* **17** 093303
[16] Lu J and Tryggavason G 2006 Numerical study of turbulent bubbly downflows in a vertical channel *Phys. Fluids* **18** 103302
[17] Lu J and Tryggavason G 2008 Effect of bubble deformability in turbulent bubbly upflow in vertical channel *Phys. Fluids* **20** 040701
[18] Brackbill J U and Ruppel H M 1986 FLIP: a method for adaptively zoned, particle-in-cell calculations in two dimensions *J. Comput. Phys.* **65** 314
[19] Charrier J, Shrivastava S and Wu R 1989 Free and constrained inflation of elastic membranes in relation to thermoforming-non-axisymmetric problems *J. Strain. Analysis* **24** 55
[20] Zinchenko A, Rother M and Davis R 1997 A novel boundary-integral algorithm for viscous interaction of deformable drops *Phys. Fluids* **9** 1493

Sound-mediated dynamic correlations between colloidal particles in a quasi-one-dimensional channel

D Frydel and H Diamant

Raymond and Beverly Sackler School of Chemistry, Tel Aviv University, Tel Aviv 69978, Israel

E-mail: hdiamant@tau.ac.il

Abstract. We study the hydrodynamic interactions between colloids suspended in a compressible fluid inside a rigid channel. Using lattice–Boltzmann simulations and a simplified hydrodynamic theory, we find that the diffusive dynamics of density perturbations (sound) in the confined fluid give rise to particle correlations of exceptionally long spatial range and algebraic temporal decay. We examine the effect of these sound-mediated correlations on two-particle dynamics and on the collective dynamics of a quasi-one-dimensional suspension.

1. Introduction

Particles embedded in an unbounded fluid are dynamically correlated by hydrodynamic interactions that decay like $1/r$ (where r denotes the separation between particles) [1]. This result is understood in terms of the fundamental solution to the steady-state Stokes equation, $\eta\nabla^2\mathbf{u} = \boldsymbol{\nabla}p - \mathbf{b}\delta(\mathbf{r})$, supplemented by the fluid-incompressibility condition $\boldsymbol{\nabla}\cdot\mathbf{u} = 0$. Here $\mathbf{u}$ is the flow velocity field, p the pressure field, $\mathbf{b}$ a point force, and η the shear viscosity. It follows from the first equation that the vorticity, $\boldsymbol{\nabla}\times\mathbf{u}$, satisfies the Laplace equation, leading in the case of an unbounded fluid to the $1/r$ slow decay of the velocity field. When particles are confined to a quasi-one-dimensional (q1D) channel, however, these long-range dynamic correlations are cut off [2], the screening being attributed to dissipation at the fluid-solid interface, i.e., to the loss of fluid momentum at the channel boundaries. The resulting screening length is comparable to the channel width.

The dissipative effect of the boundaries can be incorporated through an additional friction term in an "effective fluid" model [3, 4], $\eta\nabla^2\mathbf{u} = \rho_0\xi\mathbf{u} + \boldsymbol{\nabla}p - \mathbf{b}\delta(\mathbf{r})$. In the extra term ρ_0 is the fluid mass density and ξ an effective friction coefficient. The parameter ξ has units of inverse time and characterizes the rate of momentum loss to the boundaries. It can be estimated, therefore, as the inverse of the time it takes fluid momentum to diffuse to the channel boundary, $\xi \sim \nu/h^2$, $\nu = \eta/\rho_0$ being the kinematic shear viscosity and h the channel width. This simplified, phenomenological approach has been shown by simulations [5, 6, 7, 8] and analytical calculations [9] to correctly reproduce the qualitative behavior of confined fluids. The modified Stokes equation leads to a Helmholtz equation for the vorticity, whose regular solutions are exponentially screened functions. Yet, the friction that suppresses the vorticity (transverse fluid stress) does not similarly suppress the pressure field (longitudinal fluid stress) emanating from a locally applied force. Consequently, the application of a localized force generates a long-range pressure distribution, which may give rise to long-range flows and forces on embedded particles. For example, in a q2D suspension confined between two plates, the pressure distribution creates hydrodynamic interactions that decay like $1/r^2$, while the $1/r$ correlations due to transverse momentum transfer are cut off [10]. Thus, although the hydrodynamic interactions in the q2D system remain long-ranged, the very mechanism for correlations changes [11]. In the case of a q1D channel, the pressure generated by a localized force far along the channel is constant and, therefore, does not produce long-range forces. Thus, within the Stokes description of a steady incompressible flow, in a q1D rigid channel long-range hydrodynamic interactions arise neither from transverse stresses nor from longitudinal ones, and the correlation between particles is screened, the screening length being proportional to the channel width [2].

The omission of compressibility, and with it of compression (longitudinal sound) modes, is usually justified for unbounded fluids, where the effects of sound are short-lived. The short propagation time to any reasonable distance r, $\tau_s \sim r/c_s$, is a result of the large speed of sound, $c_s \sim 10^3$ m/s for ordinary liquids such as water. The effect of sound on the short-time hydrodynamic interactions in unconfined suspensions was addressed in earlier works [12, 13, 14]. The corresponding effects are subtle and short-lived, quickly taken over by the fully established hydrodynamic interaction due to transverse flow.

In confined geometries, however, the behavior of sound modes qualitatively changes. Inclusion of fluid compressibility has been found to lead to a velocity autocorrelation function of a confined particle, which decays with time only algebraically, with a negative long-time tail [5, 6, 7, 15]. [In the case of a q1D channel the decay is $\sim (-t^{-3/2})$.] Thus, under confinement, sound acquires long-time memory without a well-defined relaxation time, and the incompressibility assumption loses, at least in principle, its justification. Informed by this result, one suspects that the long memory of sound modes may play a role in the dynamic correlations between different particles in a channel: since sound may propagate to long distances, the resulting correlations will be long-ranged, and the algebraic temporal decay will endow these interactions with a long-time memory. This novel mechanism for correlations challenges the conventional wisdom described above, according to which fluid-mediated correlations in a rigid channel are exponentially screened. The question that comes to mind is whether the long-time memory of sound is sufficiently significant to be observed experimentally and to undermine predictions of the theories based on the assumption of incompressibility. We remind that the algebraic temporal decay of sound modes in a channel does not imply that sound is capable of causing steady rearrangement of the fluid. The integrated effect of the $t^{-3/2}$ tail decays as $t^{-1/2}$, which is very slow, but still implies that at infinite time (the steady-state limit) all perturbations due to sound eventually vanish.

In a recent Letter [16] we have demonstrated that the heuristic arguments given above are correct — namely, the diffusive, slowed-down sound modes of a fluid confined in a rigid channel mediate long-range and long-time velocity correlations between suspended particles. In the present publication we report detailed findings concerning the compressive response of the fluid and its effect on particles, which were not included in that brief publication. Two tools are employed and compared: simplified analytical calculations and lattice–Boltzmann simulations. We assume no-slip boundary conditions at the fluid-particle interface and consider either slip or no-slip boundary conditions at the channel boundaries. Fluid dynamics in a q1D channel strongly depends on the boundary conditions imposed at the edges of the channel [17]. In the calculations we assume an infinitely long channel with a vanishing flow velocity at the edges. In the simulation we take a sufficiently long channel, such that all the temporal results presented below are insensitive to the system size. Further details of the simulations can be found in Ref. [16].

2. Dynamics of fluid density perturbations

2.1. Unbounded fluid

We begin by revisiting the compressive response of an unbounded Newtonian fluid to an impulsive force [18], to which we will later compare the case of a fluid confined in a channel [15, 19].

Isothermal fluid dynamics [20] is governed by the momentum-conservation Navier–Stokes equation,

$$\rho\Big[\partial_t\mathbf{u} + (\mathbf{u}\cdot\boldsymbol{\nabla})\mathbf{u}\Big] = -\nabla p + \eta\nabla^2\mathbf{u} + (\eta_v + \eta/3)\boldsymbol{\nabla}(\boldsymbol{\nabla}\cdot\mathbf{u}) + \mathbf{f}, \tag{1}$$

and the mass-conservation equation,

$$\partial_t\rho + \boldsymbol{\nabla}\cdot(\rho\mathbf{u}) = g. \tag{2}$$

In these equations ρ is the fluid density field, $\mathbf{f}$ an external force density, and η and η_v are the shear and volume viscosities, respectively. The function g could represent, for example, a mass monopole, $m\delta(\mathbf{r}-\mathbf{r}_0)$, where at the point $\mathbf{r}_0$ fluid is generated or lost, or a mass dipole, $-\mathbf{d}\cdot\boldsymbol{\nabla}\delta(\mathbf{r}-\mathbf{r}_0)$, where fluid is at the same time created and lost, thus globally conserving mass but generating a flow.

The hydrodynamic equations are linearized by neglecting the convective term, $(\mathbf{u}\cdot\boldsymbol{\nabla})\mathbf{u}$, and separating the thermodynamic variables into equilibrium and small perturbation parts, $\rho(\mathbf{r},t) = \rho_0 + \delta\rho(\mathbf{r},t)$ and $p(\mathbf{r},t) \simeq p_0 + (\partial p/\partial\rho)\delta\rho = p_0 + c_s^2\delta\rho(\mathbf{r},t)$. The resulting linearized equations read

$$\rho_0\partial_t\mathbf{u} = -c_s^2\nabla\delta\rho + \eta\nabla^2\mathbf{u} + (\eta_v + \eta/3)\boldsymbol{\nabla}(\boldsymbol{\nabla}\cdot\mathbf{u}) + \mathbf{f}, \tag{3}$$

$$\partial_t\delta\rho + \rho_0\boldsymbol{\nabla}\cdot\mathbf{u} = g. \tag{4}$$

When dealing with sound, it is convenient to decompose the linearized equations into transverse and longitudinal components (the so called Helmholtz decomposition), $\mathbf{u} = \mathbf{u}^T + \mathbf{u}^L$, which satisfy $\boldsymbol{\nabla}\cdot\mathbf{u}^T = 0$ and $\boldsymbol{\nabla}\times\mathbf{u}^L = 0$. The decomposition yields

$$\rho_0\partial_t\mathbf{u}^T = \eta\nabla^2\mathbf{u}^T + \mathbf{f}^T, \tag{5}$$

and

$$\rho_0\partial_t\mathbf{u}^L = -c_s^2\boldsymbol{\nabla}\delta\rho + (\eta_v + 4\eta/3)\nabla^2\mathbf{u}^L + \mathbf{f}^L, \tag{6a}$$

$$\partial_t\delta\rho + \rho_0\boldsymbol{\nabla}\cdot\mathbf{u}^L = g, \tag{6b}$$

where we have used the identity $\boldsymbol{\nabla}(\boldsymbol{\nabla}\cdot\mathbf{u}) = \nabla^2\mathbf{u}^L$. After separating the force-density field into transverse and longitudinal components, $\mathbf{f} = \mathbf{f}^T + \mathbf{f}^L$, the two flows become independent. The sound is associated with the longitudinal flow, which is coupled dynamically to the density—i.e., the sound modes characterize the propagation and relaxation of a density perturbation.

The time evolution of a density perturbation is obtained by taking the divergence of Eq. (6a) and then transforming it using the mass-conservation equation, Eq. (6b),

$$\partial_t^2\delta\rho = c_s^2\nabla^2\delta\rho + 2\Gamma\nabla^2\partial_t\delta\rho - \mathbf{b}\cdot\boldsymbol{\nabla}\delta(\mathbf{r})\delta(t). \tag{7}$$

In Eq. (7) $\Gamma = \frac{1}{2}\frac{\eta_v+4/3\eta}{\rho_0}$, and we have set $g = 0$ and the force density to an impulsive point force, $\mathbf{f} = \mathbf{b}\delta(\mathbf{r})\delta(t)$. Note that the application of the impulsive point force generates a momentary mass dipole in the equation for the density, $-(\mathbf{b}\delta(t))\cdot\boldsymbol{\nabla}\delta(\mathbf{r})$. Equation (7) represents a damped wave equation with a damping parameter Γ. If $\Gamma = 0$, we retrieve a wave equation for $\delta\rho$, and if $c_s = 0$, we obtain a diffusion equation for $\partial_t \delta\rho$. Upon Fourier-transforming the spatial coordinates, Eq. (7) becomes

$$\partial_t^2 \delta\rho_k + (2\Gamma k^2)\partial_t \delta\rho_k + (c_s^2 k^2)\delta\rho_k = -i\mathbf{k}\cdot\mathbf{b}\delta(t),$$

whose solution reads

$$\delta\rho_k(t) = \frac{-i\mathbf{k}\cdot\mathbf{b}e^{-\Gamma k^2 t}}{\sqrt{c_s^2 k^2 - \Gamma^2 k^4}} \sin\left(t\sqrt{c_s^2 k^2 - \Gamma^2 k^4}\right). \tag{8}$$

If $c_s^2 - \Gamma^2 k^2 > 0$, the density perturbation propagates as a decaying wave. If, on the other hand, $c_s^2 - \Gamma^2 k^2 < 0$, the sine becomes a hyperbolic sine, and the density perturbation propagates diffusively. The crossover between these two behaviors occurs above a wavelength $\sim \Gamma/c_s$. For water $\Gamma/c_s \approx 1$ nm, while a typical colloid diameter is $\sigma \approx 10^3$ nm. Therefore, particle correlations arising from sound are mediated by essentially underdamped wave-like modes. In this large-wavelength limit, $k^{-1} \gg \Gamma/c_s$, Eq. (8) further simplifies to

$$\delta\rho_k(t) \approx \frac{-i\mathbf{k}\cdot\mathbf{b}e^{-\Gamma k^2 t}}{c_s k} \sin(c_s k t). \tag{9}$$

Figure 1 shows a snapshot of the density perturbation, generated a certain given time after the application of an impulsive force inside an unbounded fluid. We compare the result of a lattice–Boltzmann simulation, where the force is applied to a rigid spherical particle, with Eq. (8) for a point force. The analytical calculation reproduces all the features of the simulation, with a quantitative discrepancy that can be ascribed to the finite particle size in the simulation.

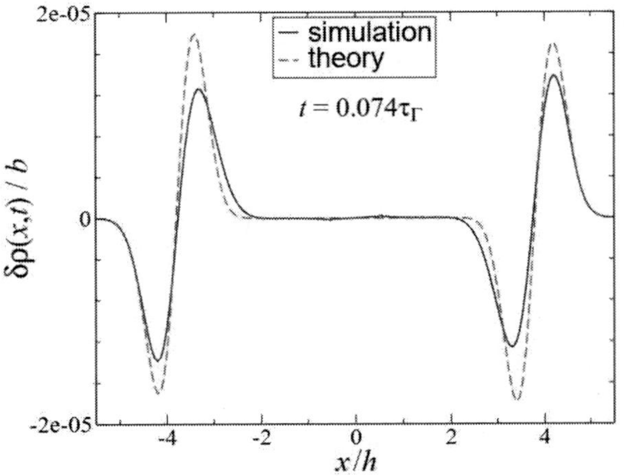

Figure 1. Snapshot of the density perturbation generated by an impulsive force in an unbounded fluid. The force is applied at the origin in the positive x direction. Lattice–Boltzmann simulation result (where the force is applied to a rigid sphere of diameter σ) is compared with the analytical one for a point force [Fourier-inverted Eq. (8)]. The snapshot is taken $0.074\tau_\Gamma$ after the impulse, where $\tau_\Gamma = h^2/\Gamma$ and $h = 1.5\sigma$.

2.2. Channel with slip boundary conditions

For a fluid confined in a channel with slip boundary conditions, we neglect variations of the density over the channel cross-section and consider only the direction along the channel, x.

Equation (7) is then rewritten as

$$\partial_t^2 \delta\rho = c_s^2 \delta\rho'' + 2\Gamma\delta\partial_t\rho'' - \frac{b}{h^2}\delta'(x)\delta(t),$$

where a prime indicates differentiation with respect to x, and an impulse is evenly distributed over the cross–sectional area of a square channel of side h. In the long wavelength limit the solution takes the form,

$$\delta\rho \approx \frac{b}{2c_s h^2}\left(\frac{e^{-(x-c_st)^2/(4\Gamma t)}}{\sqrt{4\pi\Gamma t}} - \frac{e^{-(x+c_st)^2/(4\Gamma t)}}{\sqrt{4\pi\Gamma t}}\right). \quad (10)$$

There are two signals propagating with the speed of sound in opposite directions. In addition, each signal spreads out diffusively with diffusion coefficient Γ. Figure 2 shows density perturbation snapshots generated by an impulse applied at the center and along a square channel with slip boundary conditions. Lattice–Boltzmann simulation results, where the impulse is applied to a spherical particle, agree well with Eq. (10).

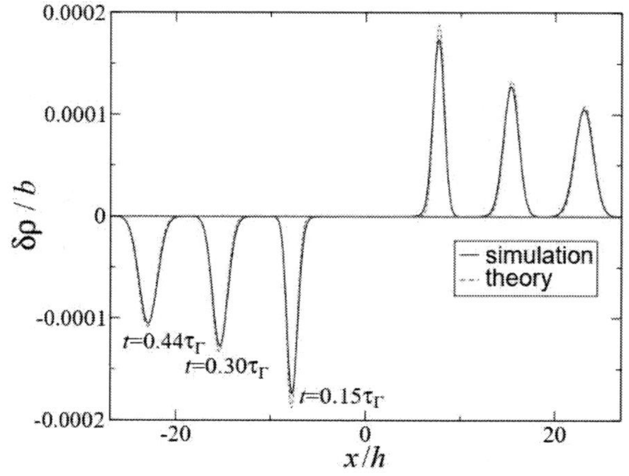

Figure 2. Three snapshots of a density perturbation for a fluid in a square channel with slip boundary conditions. An impulsive force is applied along the channel axis in the positive x direction. Lattice–Boltzmann simulation results (where the force is applied to a rigid sphere of diameter σ) are compared with the analytical ones for a force distributed uniformly over the square cross-section of side $h = 1.5\sigma$ [Eq. (10)]. The time unit is $\tau_\Gamma = h^2/\Gamma$.

2.3. Channel with no-slip boundary conditions

To account for no-slip boundary conditions at the channel walls, we introduce an effective friction term, $(-\rho_0\xi\mathbf{u}^L)$, to the right-hand side of Eq. (6a). This leads to an extra term, $(-\xi\partial_t\delta\rho)$, in Eq. (7). Thus, neglecting again density variations transverse to the channel axis, we have the effective 1D equation,

$$\partial_t^2 \delta\rho = c_s^2 \delta\rho'' + 2\Gamma\partial_t\delta\rho'' - \xi\partial_t\delta\rho - \frac{b}{h^2}\delta'(x)\delta(t).$$

For the friction coefficient (rate of momentum loss to the walls) we take $\xi = \alpha\nu/h^2$, where $\nu = \eta/\rho_0$ is the kinematic shear viscosity, and α a geometrical prefactor, dependent on the shape of the channel cross-section. (The value of α can be determined from the resistance of the channel to a steady pressure-driven flow, e.g., $\alpha \simeq 28.454$ for a square cross-section and $\alpha = 36$ for a circular one [1, 9].)

In Fourier space the solution reads

$$\delta\rho_k(t) = \frac{b}{h^2}\frac{(-ik)e^{-(\Gamma k^2+\alpha\nu/2h^2)t}}{\sqrt{c_s^2k^2-(\Gamma k^2+\alpha\nu/2h^2)^2}}\sin\left(t\sqrt{c_s^2k^2-(\Gamma k^2+\alpha\nu/2h^2)^2}\right). \tag{11}$$

The corresponding signal propagates with underdamped oscillations only for a narrow range of small wavelengths, satisfying

$$\frac{c_s}{2\Gamma}\left(1-\sqrt{1-\frac{2\Gamma}{D_s}}\right) < k < \frac{c_s}{2\Gamma}\left(1+\sqrt{1-\frac{2\Gamma}{D_s}}\right), \tag{12}$$

where $D_s = c_s^2h^2/(\alpha\nu)$ is the sound diffusion coefficient. Hence, the friction at the walls transforms the large-scale dynamics from oscillating to diffusive. In the large-wavelength limit Eq. (11) reduces to

$$\delta\rho_k(t) \approx \left(\frac{b}{\alpha\nu}\right)(-ik)e^{-D_sk^2t}, \tag{13}$$

which in real space reads,

$$\delta\rho \approx \frac{b}{\alpha\nu}\frac{2}{\sqrt{\pi}}\frac{xe^{-x^2/4D_st}}{(4D_st)^{3/2}}. \tag{14}$$

Thus, at large distances the density perturbation evolves as a diffusive dipolar signal.

Figure 3 shows snapshots of the density perturbation propagating in a square channel with no-slip boundary conditions. The wave-like propagation quickly disappears and the perturbation continues to evolve diffusively. Once again, the simplified 1D theory is found to qualitatively reproduce all the features seen in the lattice–Boltzmann simulation [21].

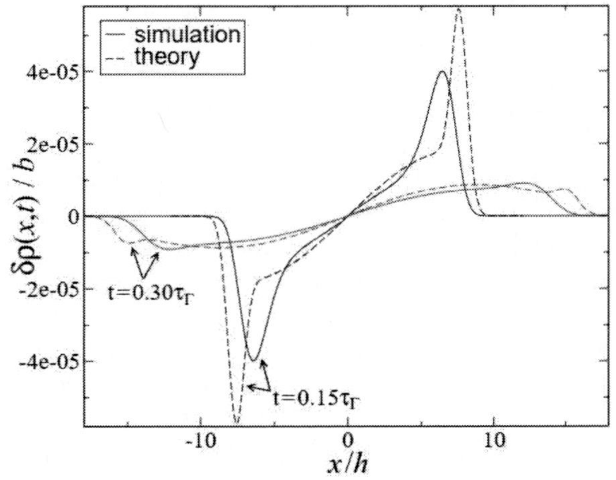

Figure 3. Two snapshots of a density perturbation for a fluid in a square channel with no-slip boundary conditions. An impulsive force is applied along the channel axis in the positive x direction. Lattice–Boltzmann simulation results (where the force is applied to a rigid sphere of diameter σ) are compared with the analytical ones for a force distributed uniformly over the square cross-section of side $h = 1.5\sigma$ (Fourier-inverted Eq. (11), multiplied by a factor of 2 [21]). The time unit is $\tau_\Gamma = h^2/\Gamma$.

3. Velocity cross-correlation function

Section 2 has been concerned with the dynamics of density perturbations in a viscous fluid. We now proceed to examine the effect of this dynamic response on particles suspended in the fluid.

Hydrodynamics represents the fluid as a continuous medium. Its discrete structure is revealed when a large particle immersed in the fluid undergoes Brownian motion due to large number of collisions with the much smaller fluid particles. Fluid particles impart momentum to the large particle, and the particle, in its turn, returns momentum to the fluid by setting off large-scale fluid flows. These flows lead to fluid-mediated correlations between the large particles. The flows can be decomposed, according to Eqs. (5) and (6), into transverse and longitudinal components. The present work is concerned with correlations caused by density perturbations, and so we focus on the role of longitudinal flows. As noted in Sec. 1, transverse flows in a rigid channel are screened at distances larger than the channel width and do not contribute to long-range correlations.

To visualize these flow-mediated correlations, we show in Fig. 4 velocity cross-correlation functions between two isolated particles, as obtained from lattice–Boltzmann simulations for three different systems: an unbounded fluid, a fluid in a square channel with slip boundary conditions, and a fluid in a square channel with no-slip boundary conditions. The cross-correlation function,

$$C(t,d) = \Big\langle V_1(0)V_2(t)\Big\rangle_d,$$

measures ensemble-averaged correlations between the velocities of two particles, V_1 and V_2, separated by a distance d. In all three cases we take the two velocities to be oriented along the line that connects the centers of the two particles. To avoid the time-consuming fluctuating simulations and averaging procedure, we employ instead a deterministic measurement of $C(t,d)$ based on the linear response of the two-particle system. The equivalence of the two measurements is guaranteed by the fluctuation-dissipation theorem [22]. The procedure goes as follows. For $t < 0$ both particles are at rest, $V_1(t) = V_2(t) = 0$. At $t = 0$ we apply an instantaneous force to particle 1 by assigning it a finite initial velocity, $V_1(t = 0) = V_0$. As time evolves, $V_2(t)$ responds to the perturbation set off by particle 1. The velocity cross-correlation function is then obtained as

$$\frac{C(t,d)}{k_BT} = \frac{1}{M}\frac{V_2(t)}{V_1(0)},$$

where M is the particle mass and k_BT the thermal energy.

In Fig. 4 we see that in all three systems there is a finite incipient time required for the signal to reach particle 2. This time is roughly d/c_s. The leading signal in the channels is larger than that in the unbounded fluid, because the perturbation is guided along the channel rather than spreading in 3D. At larger inter-particle distances, however, that signal is suppressed in the case of a rigid channel with no-slip boundary conditions because of the diffusive nature of sound. (See Fig. 5.)

The correlations at long times are very different for the three systems. In the unbounded case the correlation is governed by a positive algebraic tail, $\sim t^{-3/2}$. This is a manifestation of the well known long-time tails arising from the 3D diffusion of vorticity (monopolar transverse flow) [22]. In the rigid (no-slip) channel correlations due to transverse flows are screened, and the long-time correlation is governed by a negative algebraic tail, $\sim (-t^{-3/2})$. This is a result of the 1D diffusion of sound (dipolar longitudinal flow) presented in Sec. 2. By contrast, in the case of a channel with slip boundary conditions, the longitudinal signal propagates acoustically, making the correlation vanish exponentially fast once the signal has passed particle 2.

We next focus on the sound-mediated correlations in a rigid channel with no-slip boundary conditions. Figure 5 shows $C(t,d)$ for different separations d. The structure is similar for the different separations. There is the incipient time required for the leading signal to reach the other

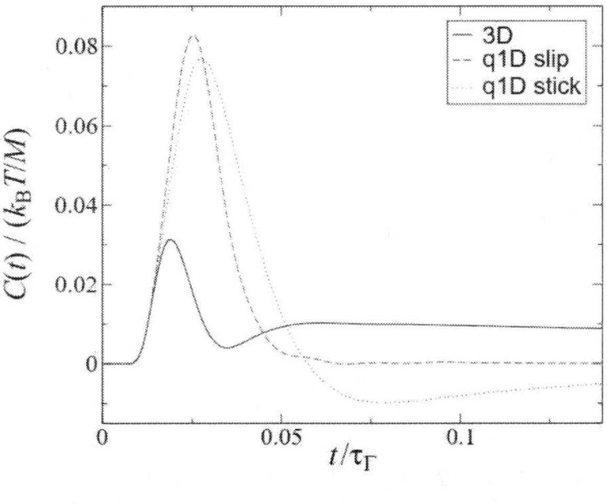

Figure 4. Velocity cross-correlation functions for two particles, as obtained from lattice–Boltzmann simulations of three systems: an unbounded fluid, a fluid in a square channel (side h) with slip boundary conditions, and a fluid in a square channel with no-slip boundary conditions. The particles have diameter σ and mass M and they are separated by a distance $d = 2\sigma = 1.33h$. The time unit is $\tau_\Gamma = h^2/\Gamma$.

particle. This signal is suppressed with increased separation because of sound diffusion. It is followed by a long-time negative algebraic tail, whose magnitude does not show any dependence on separation. On the one hand, this indicates a correlation of *unrestricted spatial range*. On the other hand, the backflow associated with the negative tail causes the particle to return almost exactly to its initial position; the time-integrated correlation (which is equal to the coupling diffusion coefficient of the particle pair) decays exponentially with d/h. This is how the two phenomena — long-range temporal correlations and short-range steady-state correlations — are reconciled.

To quantitatively account for the long-time algebraic behavior of the correlations we need to relate the diffusive density perturbation, given by Eq. (14), to the flow velocity. This can be done either via the continuity equation, $\partial_t \delta\rho = -\rho_0 \partial_x u$, or by applying Fick's law to the sound diffusion, $\rho_0 u = -D_s \partial_x \delta\rho$. Both methods yield the same long-time flow velocity,

$$u(x,t) = -\frac{b}{4c_s h \rho_0 \sqrt{\pi\alpha\nu}} \left(\frac{1}{t}\right)^{3/2}.$$

Equating it with the velocity advecting particle 2 and applying the fluctuation-dissipation theorem, as explained above, we obtain the long-time velocity cross-correlation function [23],

$$C(t,d) = -\frac{k_B T}{4c_s h \rho_0 \sqrt{\pi\alpha\nu}} \left(\frac{1}{t}\right)^{3/2}, \qquad (15)$$

which has no dependence on the inter-particle separation d. The negative sign of the long-time tail is traced back to the dipolar shape of the diffusive density perturbation, Eq. (14).

4. Collective dynamics

Finally, we study the effect of sound on the collective dynamics of many particles. The discussion in this section is restricted to the q1D geometry of a rigid channel with no-slip boundary conditions.

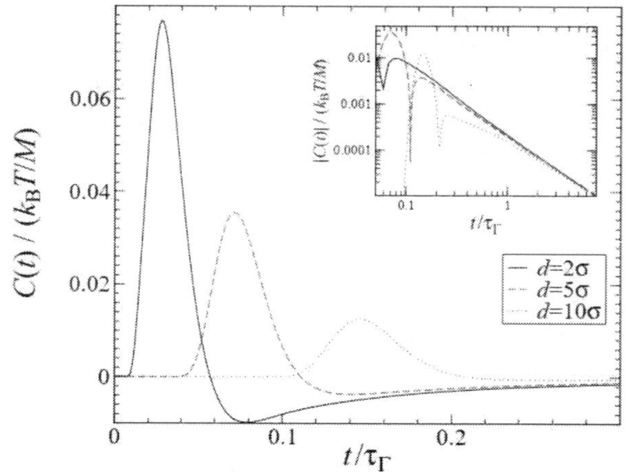

Figure 5. Velocity cross-correlation functions for two particles in a rigid channel (no-slip boundary conditions), as obtained from lattice–Boltzmann simulations. The particles, situated on the channel axis, have diameter σ and mass M. The channel has a square cross-section of side $h = 1.5\sigma$. Correlations are shown for three different inter-particle separations. The time unit is $\tau_\Gamma = h^2/\Gamma$. The inset shows the absolute value of the correlations on a logarithmic scale, demonstrating the $(-3/2)$ power law at long times, with a distance-independent amplitude.

We consider the wavenumber-dependent particle current correlation function [22],

$$J(k,t) = \frac{1}{N}\sum_{i,j}^{N}\left\langle V_i(0)V_j(t)e^{ik(x_i(0)-x_j(t))}\right\rangle, \tag{16}$$

where N is the number of particles and i, j are particle labels. Self-contributions to the sum, with $i = j$, are related to the velocity autocorrelation function of single particles, $\langle V(0)V(t)\rangle$. As our interest here lies solely in cross-correlations, we subtract this contribution and consider $J_c(k,t) = J(k,t) - \langle V(0)V(t)\rangle$.

To obtain J_c from lattice–Boltzmann simulations we assume time-scale separation between the relaxation of the fluid and that of particle configurations. We generate configurations X^N of N particles using a Monte Carlo scheme. For each configuration we assign to each particle an initial velocity from a Maxwell–Boltzmann distribution and then follow the time evolution of every particle in a lattice–Boltzmann simulation. This allows us to calculate, for each configuration, the velocity cross-correlation functions between all pairs of particles, $\langle V_i(0)V_j(t)\rangle\{X^N\}$, as described in Sec. 3. The assumption of time-scale separation simplifies the expression for J_c to

$$J_c(k,t) = \frac{\phi}{\sigma}\int_{-\infty}^{\infty} dx\, g(x)e^{ikx}\Big\langle V_1(0)V_2(t)\Big\rangle_{\text{fast}}\{x\},$$

where $g(x)$ is the equilibrium pair correlation function of the particles, and $\phi = N\sigma/L$ (L being the channel length) is their linear fraction. The angular brackets $\langle\ldots\rangle_{\text{fast}}$ denote a double averaging procedure: first, for a given configuration, an average is taken over time that is sufficiently long for the fluid to relax but sufficiently short for the configuration to be considered "frozen"; then another averaging is performed over many particle configurations with the constraint $x_1 - x_2 = x$.

In Fig. 6 we plot the time evolution of $J_c(k,t)$, thus obtained from the lattice–Boltzmann simulations, for various wavenumbers k. As $t \to 0$, J_c vanishes, as the inter-particle correlations in the compressible fluid require a finite time to develop. The presence of oscillations for small wavelengths, comparable to the particle size and channel width, reflects the wave-like propagation of sound over these small length scales. The oscillations vanish at longer wavelenghs, indicating the onset of diffusive sound propagation. At long times, the particle current correlation J_c decays algebraically as $t^{-3/2}$.

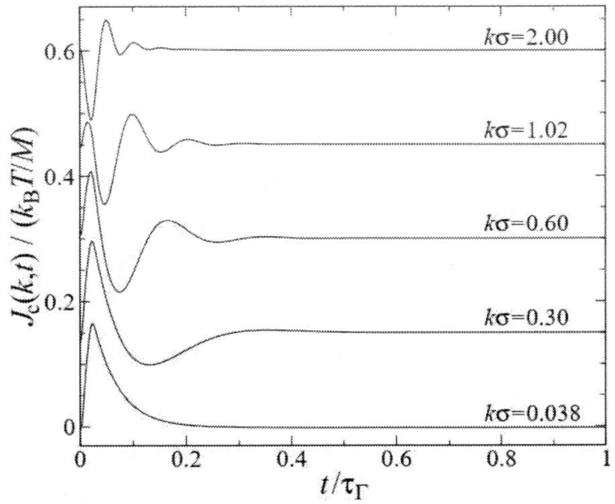

Figure 6. Particle current correlation functions obtained from lattice–Boltzmann simulations for a suspension of particles (diameter σ, linear fraction $\phi = 0.72$) in a square channel (side $h = 1.5\sigma$). The different curves, corresponding to different wavenumber k, are shifted vertically by 0.15 from each other, for clarity. The time unit is $\tau_\Gamma = h^2/\Gamma$.

To further clarify the results of Fig. 6 we resort to a simple model, where $g(x)$ is replaced by its dilute limit: $g(x) = 0$ for $|x| < \sigma$, and $g(x) = 1$ for $|x| > \sigma$. For the velocity cross-correlation function $\langle V_1(0)V_2(t)\rangle_{\text{fast}}\{x\}$ we use the 1D description of Sec. 3. From Eq. (11) and the continuity equation, $\partial_t \delta\rho_k = -ik\rho_0 u_k$, we obtain the flow velocity u_k. The fluctuation-dissipation theorem is then used to relate the flow to the velocity cross-correlations. This model yields

$$\frac{J_c(k,t)}{k_B T} = \frac{\phi}{b\sigma}\left[u_k(t) - \int_{-\sigma}^{\sigma} dx\, e^{-ikx} u(x,t)\right]. \qquad (17)$$

The qualitative change in J_c as its oscillations vanish at long wavelengths is reproduced by the first term. The onset of the overdamped diffusive regime for small wavenumbers, lying outside the range delineated in Eq. (12), should occur for the parameters used in the simulation at $k\sigma < 0.18$, which is in line with Fig. 6. The second term in Eq. (17) supplies the function J_c with the algebraic decay $\sim t^{-3/2}$, which at long wavelength is positive.

Another function that characterizes the collective dynamics of particles is the dynamic structure factor [22, 24],

$$S(k,t) = \frac{1}{N}\sum_{i,j}^{N}\left\langle e^{ik(x_i(0)-x_j(t))}\right\rangle. \qquad (18)$$

At times much shorter than the configurational relaxation time it decays exponentially with time and can be written as

$$S(k,t) = S_0(k)e^{-tk^2 H(k)/S_0(k)},$$

where $S_0(k) = S(k, t = 0)$ is the static structure factor. The hydrodynamic factor $H(k)$ accounts for the effect of fluid-mediated correlations on the dynamic structure and is given by [24]

$$H(k) = D_{\text{self}} + \frac{1}{N}\sum_{i\neq j}^{N}\left\langle D_{ij}\{X^N\}\, e^{ik(x_i - x_j)}\right\rangle,$$

where $D_{ij}\{X^N\}$ are the configuration-dependent pair diffusion coefficients, and $D_{\text{self}} = \langle D_{ii}\rangle$ is the self-diffusion coefficient. Note that $H(k)$ depends only on steady-state properties of the fluid and, hence, is unaffected by sound.

Recalling the Green–Kubo relation [22],

$$D_{ij}(x) = \int_0^\infty dt \left\langle V_i(0)V_j(t)\right\rangle\{x\},$$

and assuming time-scale separation, we relate $H(k)$ to the current correlation function,

$$H(k) = \int_0^\infty dt J(k, t).$$

To examine the effect of sound we therefore use the functions

$$\hat{H}(k, t) = \int_0^t dt'\, J(k, t'), \quad \hat{H}_c(k, t) = \int_0^t dt'\, J_c(k, t'),$$

such that in the limit $t \to \infty$ $\hat{H}(k, t)$ reduces to $H(k)$, and $\hat{H}_c(k, t)$ to $H(k) - D_{\text{self}}$ [25]. Figure 7 shows the dependence of $\hat{H}(k, t)$ on k at various times. As t increases, $\hat{H}(k, t)$ converges to its limiting form, $H(k)$, and eventually all sound effects vanish. However, at any finite time there appears a feature that is absent in unbounded suspensions—a sharp peak at $k \to 0$.

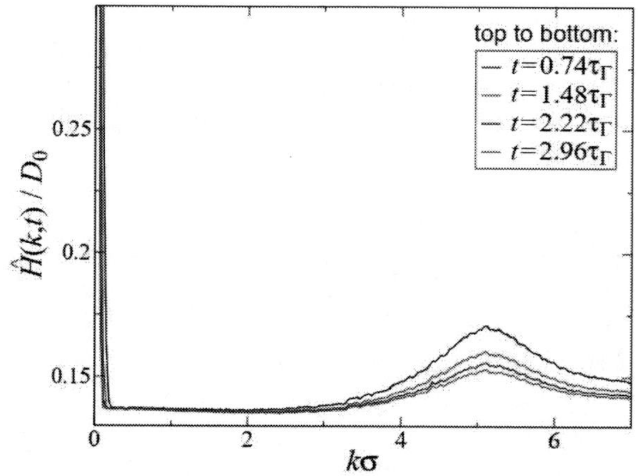

Figure 7. Temporal hydrodynamic factor as a function of wavenumber at various times. The system is the same as in Fig. 6.

To see that the small k feature is due to long-range sound-mediated correlations, we return to the simplified 1D description. Using Eq. (17) and the continuity equation, $\partial_t \delta\rho_k = -ik\rho_0 u_k$, we obtain

$$\frac{\hat{H}_c(k, t)}{k_B T} = \frac{\phi}{b\sigma\rho_0}\left[\frac{\delta\rho_k(t)}{(-ik)} - \frac{1}{\pi}\int_{-\infty}^{\infty} dk' \frac{\sin(\sigma(k' - k))}{k' - k}\frac{\delta\rho_{k'}(t)}{(-ik')}\right], \tag{19}$$

where $\delta\rho_k$ is given by Eq. (11) [or by Eq. (13) for small k]. At $t \to \infty$ this expression vanishes, indicating that within the simplified 1D description all steady correlations between particles confined in the channel disappear, and $H(k) = D_{\text{self}}$. However, at any finite t, we have a range of small wavenumbers, $k^2 < (D_s t)^{-1}$, for which Eq. (19) reduces to

$$\frac{\hat{H}_c(k,t)}{D_0} \approx \frac{3\pi\phi}{\alpha}\left(e^{-D_s k^2 t} - \frac{\sigma}{\sqrt{\pi D_s t}}\right), \tag{20}$$

where $D_0 = k_B T/(3\pi\eta\sigma)$ is the Stokes-Einstein self-diffusion coefficient in an unbounded fluid. Equation (20) contains two interesting terms originating from sound diffusion. The first contributes at $k \to 0$ a *time-independent* constant. The second adds a k*-independent* negative long-time algebraic tail, $\sim (-t^{-1/2})$, which derives from the time-integrated $t^{-3/2}$ tail of the current correlation function $J_c(k,t)$.

In Fig. 8 we replot the function $\hat{H}(k,t)$, focusing on the small k region, and compare it with the fit $\frac{\hat{H}(k,t)}{D_0} = \frac{6\pi\phi}{\alpha}e^{-D_s k^2 t}$ [21]. The agreement between theory and simulation is satisfactory, both exhibiting the small-k peak.

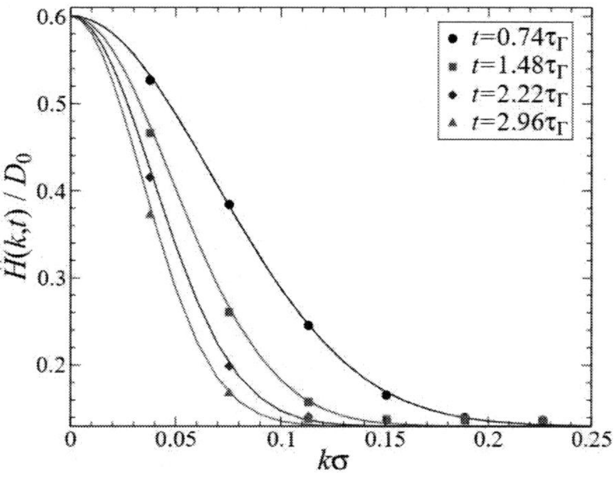

Figure 8. Temporal hydrodynamic factor for small wavenumber at various times. The system is the same as in Fig. 6. Data points represent lattice–Boltzmann simulation results. Solid lines show the expression $\frac{\hat{H}(k,t)}{D_0} = \frac{6\pi\phi}{\alpha}e^{-D_s k^2 t}$.

5. Conclusion

In the present work we have tried to establish the role of fluid density perturbations (sound) in mediating dynamic correlations among particles in a channel at the pair and the collective levels. We have been motivated in this investigation by two considerations: (i) the hydrodynamic effects that are not associated with sound (due to momentum and steady pressure) are screened in a rigid channel and have no bearing on long-range particle correlations; (ii) the density perturbations in a channel do not decay within a well defined relaxation time (as in an unbounded fluid) but rather relax algebraically, thus giving sound a long memory.

The effects of sound on particle correlations in a rigid channel have been found to be quite striking. At the two-particle level, the velocity cross-correlation function is independent of inter-particle distance (i.e., unrestricted in its range) and has a long-time negative algebraic tail $\sim (-t^{-3/2})$ (similar to the single-particle autocorrelation function [5, 6, 7]). At the collective level of a large particle assembly (q1D suspension), the particle current correlation function has a similar long-time algebraic tail. We have characterized the collective correlations by a time-dependent hydrodynamic factor, $\hat{H}_c(k,t)$, which converges to the ordinary hydrodynamic factor at infinite time [25]. The sound-mediated correlations in the channel make this function exhibit an anomalous peak at small wavenumbers and a long-time algebraic temporal decay $\sim(-t^{-1/2})$. Thus, in summary, the common picture of negligible, exponentially screened hydrodynamic interactions among particles in a rigid channel is somewhat misleading; strictly speaking, it applies only at steady state.

At the same time, the predicted sound-mediated correlations are weak and may be measurable only over sufficiently short times. The correlation is inversely proportional to the speed of sound [see, e.g., Eq. (15)], which makes its amplitude small, and the sound diffusion is fast — for water in a micron-wide channel, $D_s = c_s^2h^2/(\alpha\nu) \sim 0.1$ m^2/s. For example, let us examine the correlation between the observable Brownian displacements of two distant particles along the channel, as characterized by their pair diffusivity,

$$D_{12}(t) \equiv \frac{\langle \Delta x_1 \Delta x_2 \rangle}{2t} = \frac{1}{t}\int_0^t dt' \int_0^{t'} dt'' C(t'',d) = \frac{k_BT}{c_sh\rho_0\sqrt{\pi\alpha\nu}}t^{-1/2}. \quad (21)$$

For a micron-wide channel filled with water, according to Eq. (21), measuring D_{12} of order 10^{-3} μm^2/s would require a temporal resolution of order 10^{-7} s. Thus, the predicted sound-mediated effects might be observable using experimental techniques of high temporal and spatial resolution, such as diffusive wave spectroscopy [12] or fast optical tracking [14, 26]. The fact that D_{12} is independent of inter-particle distance may greatly facilitate the acquisition of statistics in such fast-tracking experiments.

The short-time relevance of sound-mediated correlations was noted in earlier works on unconfined suspensions [12, 13, 14]. We end by underlining the key differences between those correlations and the ones in a rigid channel addressed here. First, whereas in an unconfined suspension sound propagates as an underdamped wave until it is scattered by a sufficient number of particles [12], in a channel, because of scattering from the boundaries, sound becomes diffusive as soon as the distance gets much larger than the channel width. Second, in an unconfined suspension sound-mediated correlations are a sub-dominant effect, quickly taken over by vorticity diffusion (transverse modes) [12, 13]. By contrast, in a rigid channel correlations due to transverse flow are cut off, leaving longitudinal flow as the sole mechanism of long-range correlations. Thus, any correlation which could be resolved at inter-particle distances much larger than the channel width should be associated with sound-mediated effects.

Acknowledgments
This research has been supported by the Israel Science Foundation (Grants No. 588/06 and No. 8/10).

References and Notes
[1] Happel J and Brenner H 1983 *Low Reynolds Number Hydrodynamics* (The Hague: Martinus Nijhoff)
[2] Cui B, Diamant H and Lin B 2002 *Phys. Rev. Lett.* **89** 188302
[3] Brinkman H C 1949 *Appl. Sci. Res. A* **1** 27
[4] Ramaswamy S and Mazenko G F 1982 *Phys. Rev. A* **26** 1735
[5] Hagen M H J, Pagonabarraga I, Lowe C P and Frenkel D 1997 *Phys. Rev. Lett.* **78** 3785
[6] Pagonabarraga I, Hagen M H J, Lowe C P and Frenkel D 1998 *Phys. Rev. E* **58** 7228
[7] Pagonabarraga I, Hagen M H J, Lowe C P and Frenkel D 1999 *Phys. Rev. E* **59** 4458
[8] Frydel D and Rice S A 2007 *Phys. Rev. E* **76** 061404
[9] Felderhof B U 2011 *J. Chem. Phys.* **134** 024505
[10] Cui B, Diamant H, Lin B and Rice S A 2004 *Phys. Rev. Lett.* **92** 258301
[11] Diamant H 2009 *J. Phys. Soc. Jpn.* **78** 041002
[12] Ladd A J C, Gang H, Zhu J X and Weitz D A 1995 *Phys. Rev. E* **52** 6550

[13] Bakker A F and Lowe C P 2002 *J. Chem. Phys.* **116** 5867
[14] Henderson S, Mitchell S and Bartlett P 2002 *Phys. Rev. Lett.* **88** 088302
[15] Felderhof B U 2010 *J. Fluid Mech.* **644** 97
[16] Frydel D and Diamant H 2010 *Phys. Rev. Lett.* **104** 248302
[17] Navardi S and Bhattacharya S 2010 *J. Math. Phys.* **51** 043102
[18] Felderhof B U 2010 *Physica A* **389** 5602
[19] Felderhof B U 2011 *J. Fluid Mech.* **668** 100
[20] Landau L D and Lifshitz E M 1987 *Fluid Mechanics* 2nd edition (Pergamon: Oxford)
[21] The fields (density, flow velocity) in the 1D model correspond to cross-sectional averages of the actual fields, whereas in the simulation we measure quantities at the center of the channel. This makes the simulation result larger than the analytical one by a factor of ≈ 2. (For a parabolic velocity profile in a circular tube the cross-sectional average is exactly $1/2$ the central value [9].)
[22] Hansen J P and McDonald I R 1986 *Theory of Simple Liquids* 2nd edition (London: Elsevier)
[23] Because of cross-sectional averaging [21], this result of the 1D model is smaller than the corresponding long-time behavior in the simulations by a factor of ≈ 4. The additional factor of ≈ 2 arises from the continuity equation (or Fick's law), relating the cross-sectional averages of density (a roughly uniform transverse profile) and velocity (a roughly parabolic profile).
[24] Pusey P N 1991 in *Liquids, Freezing and Glass Transition II* eds J P Hansen, D Levesque and J Zinn-Justin (Amsterdam: Elsevier) pp 763–942
[25] A time-dependent hydrodynamic factor was introduced before, using a different definition [12]. The relation between our function, $\hat{H}_c^{(2)}$, and the one used in Ref. [12], $\hat{H}_c^{(1)}$, is $\hat{H}_c^{(1)}(k,t) = t^{-1}\int_0^t dt' \hat{H}_c^{(2)}(k,t')$.
[26] Huang R, Chavez I, Taute K M, Lukić B, Jeney S, Raizen M G and Florin E-L 2011 *Nat. Phys.* **7** 576

The swapping-trajectory effect: lattice evolution and buckling transition in wall-bounded hydrodynamic crystals

Jerzy Blawzdziewicz[1] **and Eligiusz Wajnryb**[2]

[1]Department of Mechanical Engineering, Texas Tech University, Lubbock, TX 79409, U.S.A.

[2]Institute of Fundamental Technological Research, Warsaw, Polish Academy of Sciences

E-mail: `jerzy.blawzdziewicz@ttu.edu`

Abstract.
We analyze novel structural transformations in perturbed periodic square monolayers of microspheres in parabolic flow between two parallel walls. We find that a perturbed monolayer is initially stabilized by the swapping-trajectory mechanism that causes the particles to fluctuate between faster and slower streamlines in such a way that particle collisions do not occur. The fluctuations slowly decay in time, and the array achieves nearly perfect rectangular order. Surprisingly, after the fluctuations have dissipated, the particle lattice undergoes a sudden buckling instability that produces coherent vertical displacements of particle rows oriented in the flow direction. The instability results in formation of a disordered phase in which particles are arranged into meandering strings, similar to the structures observed in recent experiments [2012 PNAS **109** 63]. We show that the behavior of the system is controlled by the swapping-trajectory interactions at all stages of the evolution.

1. Introduction

Collective hydrodynamic phenomena in microconfined suspensions figure prominently in many scientific fields ranging from dynamics of living cells and active matter [1–3] to microfluidics [4–6]. A number of recent studies [5–15] have demonstrated an unexpectedly complex phenomenology of strongly confined dispersion flows, in systems where the channel thickness is comparable to the particle size. Such geometry is frequently encountered in microfluidic devices, so strongly confined dispersions are not only of fundamental but also of practical interest (e.g., for separation of particles of different properties [16]).

Even though suspensions and emulsions in pores and narrow channels are common in natural and engineered systems, distinct phenomenology of strongly confined dispersion flows has been recognized only recently. It has been shown that particle interactions with confining walls can produce macroscopic effects such as intrinsic convection [17–19], enhanced hydrodynamic diffusion [20, 21], complex instabilities of particle jets [22], and enhanced breakup of suspension drops [23]. Particle-wall interactions can also cause microstructural changes, e.g., suspension layering [24] and development of complex micro-scale patterns in confined emulsion [4, 11, 25, 26] and suspension [12, 27] flows.

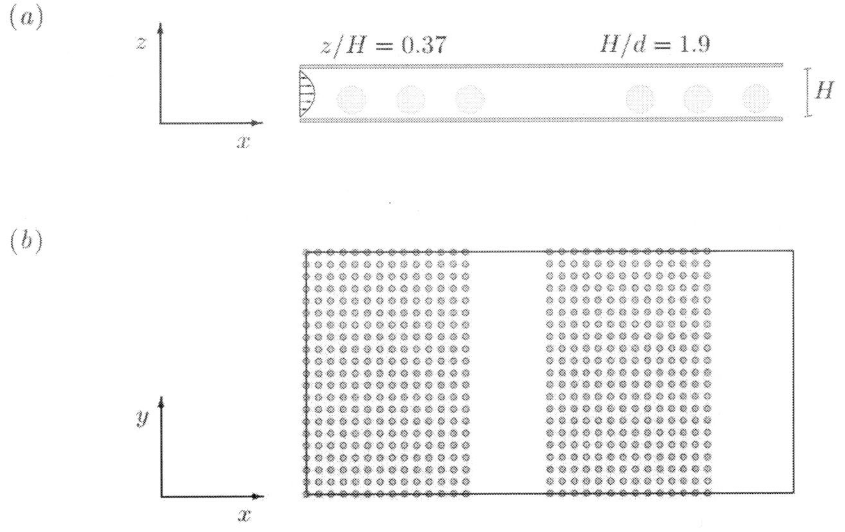

Figure 1. System definition: a square array of spheres of diameter d in a parallel-wall channel. (a) System geometry (side view); (b) initial condition (top view).

Our paper focuses on the behavior of non-Brownian rigid particles under creeping-flow conditions in parallel-wall channels with wall separation comparable to the particle diameter. Wall-induced collective evolution that is observed in such systems includes propagation of dissipative displacement waves in particle or drop trains [6, 10, 12], topological transitions in quasi-two-dimensional (2D) monolayers of microspheres [12], and unusual stability of square particle lattice [12, 28].

Our recent simulations [28] of an initially ordered particle monolayer in confined Poiseuille flow have revealed other striking and unexpected hydrodynamic phenomena, such as stabilization of an ordered particle monolayer by random fluctuations, spontaneous decay of such fluctuations with time, and a sudden instability of the particle lattice that happens right after the fluctuations have disappeared. This instability results in a new suspension microstructure with particles arranged into meandering strings. In the present study we elucidate the nature of these phenomena and discuss them in detail.

2. Evolution of a particle monolayer

We consider evolution of an infinite array of spheres of diameter d in Poiseuille flow in a parallel-wall channel of width H. The walls are in the planes $z = 0$ and $z = H$ (where z is the transverse coordinate), and the flow is in the x-direction. As depicted in Fig. 1, the particles are initially arranged on a square lattice, with the lattice vectors along the flow direction x and vorticity direction y, in an off-center plane parallel to the walls. The lattice is organized into uniform parallel strips separated by particle-free gaps. Periodic boundary conditions are applied in the x and y directions.

The simulations presented in this paper were performed for the following system parameters: the normalized channel width is $H/d = 1.9$, the initial normalized interparticle distance is $W/d = 2$, and the initial position of the particle monolayer is $z/H = 0.37$. We note, however, that a similar behavior to the one described below has been observed also for other parameter values. In what follows, the dimensionless time t is normalized by the time in which an isolated particle in the reference plane $z/H = 0.37$ travels the distance equal to its diameter.

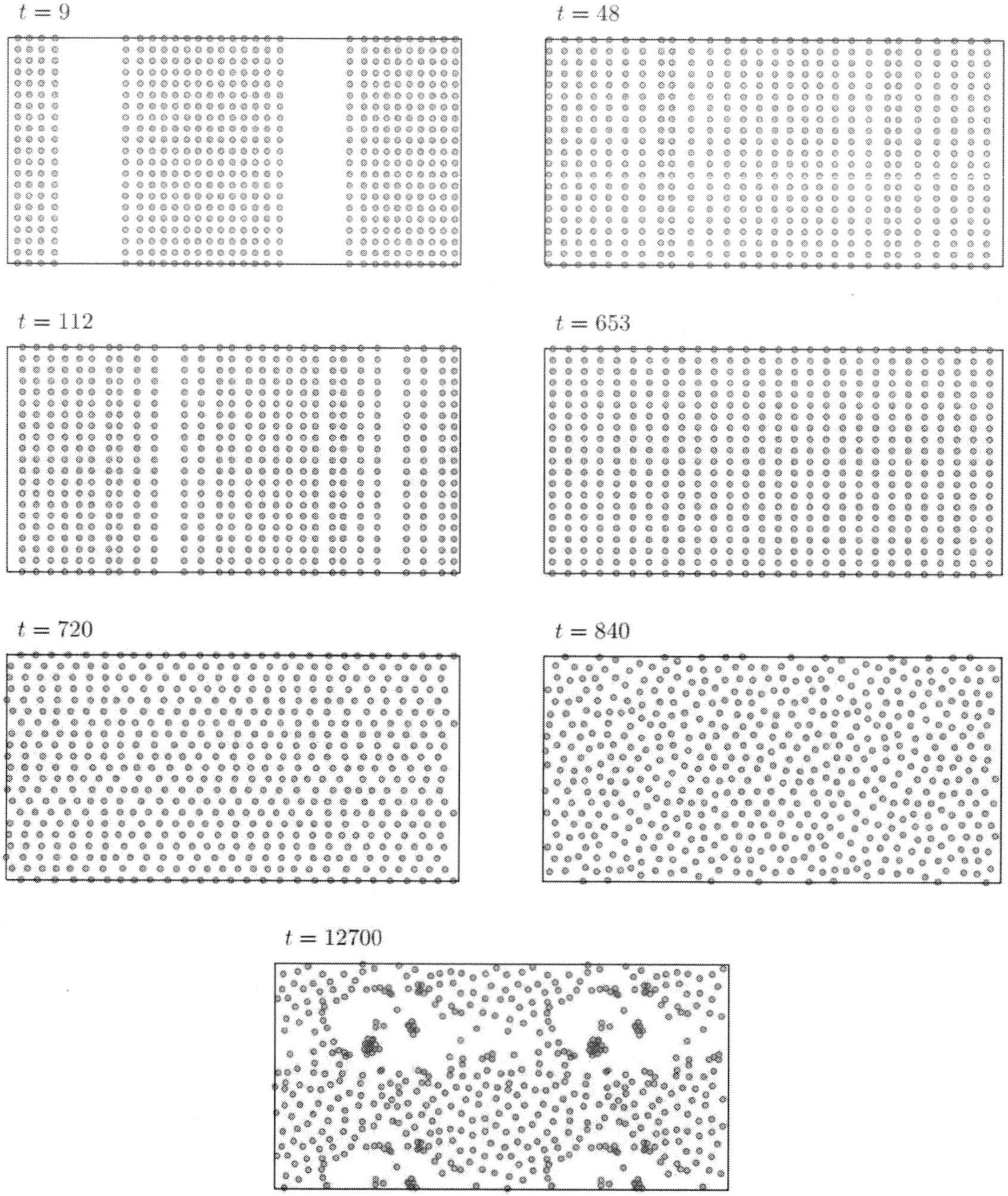

Figure 2. Evolution of a particle monolayer (top view). The panels show the projection of the particle outlines onto the x–y plane at time as labeled. Simulation cell is 14×14 particles; two periodic cells are shown in the x direction and 1.5 periodic cell in the y direction. (Also see supplemental video movie_1.mov.)

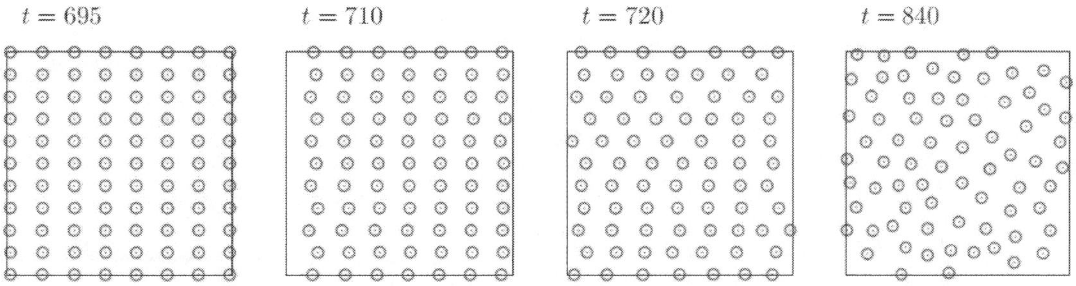

Figure 3. The lattice instability event (top view). The panels show the projection of the particle outlines onto the x–y plane at times as labeled. (Also see supplemental video movie_2.mov.)

The simulations in this paper were performed using our Cartesian representation method [29], which combines the HYDROMULTIPOLE technique [30] with expansion of the flow field in the wall presence into lateral Fourier modes [29]. The periodic boundary conditions were implemented using the far-field Hele–Shaw form of the flow in the channel [31, 32].

2.1. Microstructural evolution

The dynamics of the microstructure of the particle monolayer are depicted in Figs. 2 and 3, which show top view of the suspension in the channel (i.e., a view in the x–y plane). The system exhibits several distinct stages of evolution, and it undergoes a sudden microstructural transition. The key dynamical events, illustrated in Fig. 2 and 3, can be characterized as follows:

(i) At the first stage of evolution (cf. frames $t = 9$ and $t = 48$), individual strips of the particle lattice expand, until they completely fill the particle-free gaps, and the system forms an approximately uniform rectangular lattice.

(ii) At the second stage (cf. frames $t = 112$ and $t = 653$), the distance between y-rows in the particle lattice fluctuates significantly, but the particles remain well aligned in the y-direction, and the lattice preserves its integrity. The magnitude of the fluctuations slowly decreases in time, and at $t \approx 650$ the fluctuations nearly completely disappear.

(iii) At the third stage of evolution, when the fluctuations have decayed and a nearly perfect uniform lattice has been formed, the lattice undergoes a sudden instability (described in Sec. 2.3) and the particle monolayer exhibits coexisting local regions with square and hexagonal symmetry (cf. frame $t = 720$). As depicted in Fig. 3, the instability develops over a short time period $\Delta t \approx 10$, and afterwards the system evolves further into a new stable disordered microstructure.

(iv) At the fourth stage (cf. frame $t = 840$), the monolayer forms a complex stationary microstructure with particles arranged into meandering strings whose orientations vary between the flow and the vorticity directions. Our numerical simulations show that the system can remain in this phase for a very long time.

(v) At very long times (cf. frame at $t = 12700$) the suspension randomly undergoes another instability, resulting in formation of particle clusters that move slower than individual particles in the monolayer. The clusters gradually grow, creating particle-depleted void regions around them.

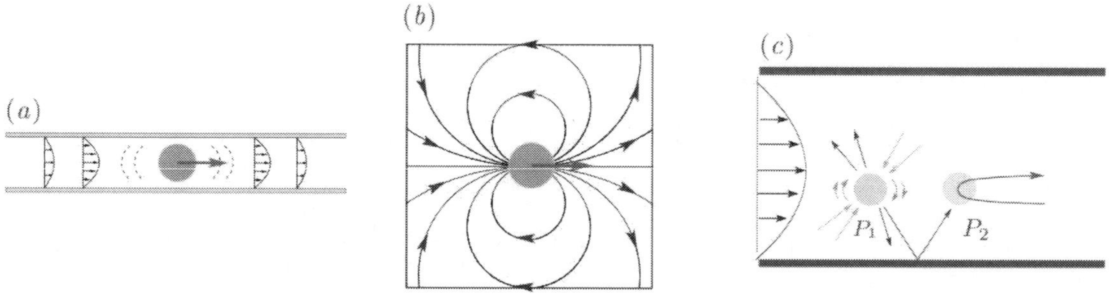

Figure 4. (*a*) Near-field and far-field flow produced by a particle moving in a channel: far from the particle (compared to the wall separation) the flow field assumes the parabolic Hele–Shaw form (side view). (*b*) The *backflow pattern* associated with 2D Hele–Shaw pressure dipole produced by a particle in an external flow (top view). (*c*) The *swapping-trajectory effect*: wall reflection of scattered flow produced by particle P_1 makes particle P_2 migrate cross-streamline, from slow moving to fast moving fluid region. As a result of this effect, particles that initially approach each other separate without collision, after switching (swapping) their cross-streamline positions (side view).

2.2. Hele–Shaw dipoles and swapping-trajectory mechanism

We argue that the complex evolution of a suspension monolayer described in Sec. 2.1 stems from two wall-induced hydrodynamic phenomena, schematically illustrated in Fig. 4. The first effect is due to the far-field Hele–Shaw dipolar flow produced by the particles [7, 33] at distances exceeding the wall separation [cf. Fig. 4(*b*)], and the second is the swapping-trajectory (ST) effect [21, 24, 34] associated with near-field wall-mediated scattered flow [cf. Fig. 4(*c*)]. While for pressure-driven particle arrays the dipolar flow affects both center-plane and off-center particle positions, the ST mechanism influences only particles that are in an off-center configuration in a local shear flow.

The role of interparticle hydrodynamic interactions associated with dipolar scattered flow has been described in our previous studies of particle monolayers moving in the midplane of a channel [12, 28]. We have shown that dipolar interactions stabilize square arrangements of hydrodynamically coupled particles, and such arrangements can withstand significant deformations (while maintaining the particle alignment). The dipolar scattered flow plays a similar stabilizing role in the particle configuration considered here. Therefore, since this phenomenon has already been well documented, in what follows we focus on the ST mechanism, which in Poiseuille flow occurs for asymmetric off-center configurations.

As illustrated in Fig. 4(*c*), the ST mechanism originates from the scattering of local incoming shear flow by one of the particles [21]. The scattered flow is then reflected from a wall and acts on the other particle, displacing it across streamlines to a fluid region moving with a different velocity. The ST phenomenon is present both in two-particle and multiparticle systems. In the case of a particle chain aligned in the flow direction (cf. Fig. 5), the ST effect causes the trailing particle to move towards the closer wall while the leading particle moves away from this wall. Since the flow is slower near the wall, the trailing particle stays behind, and the leading particle accelerates. As a result of this behavior the particle chain expands.[1] We demonstrate below that entire strips of particle lattice depicted in Fig. 2 are affected by a similar ST mechanism.

The influence of the ST effect on the motion of the particle monolayer described in Sec. 2.1

[1] We note that chain expansion can also be produced by dipolar interactions (for particles in the midplane) [35], but in our off-center system the dominant effect comes from the ST mechanism.

Journal of Physics: Conference Series **392** (2012) 012008 doi:10.1088/1742-6596/392/1/012008

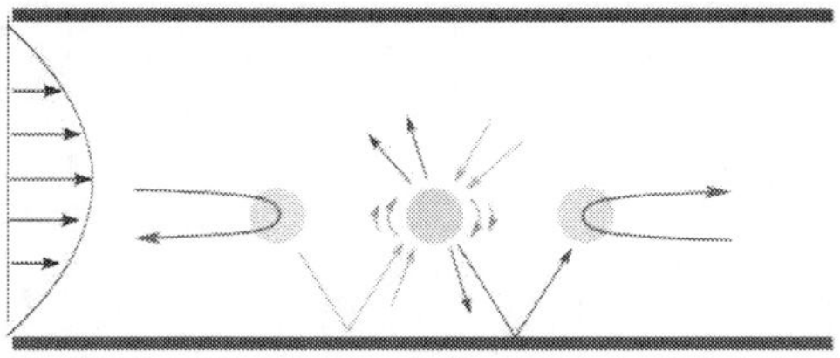

Figure 5. Swapping-trajectory effect in a three-particle system (side view).

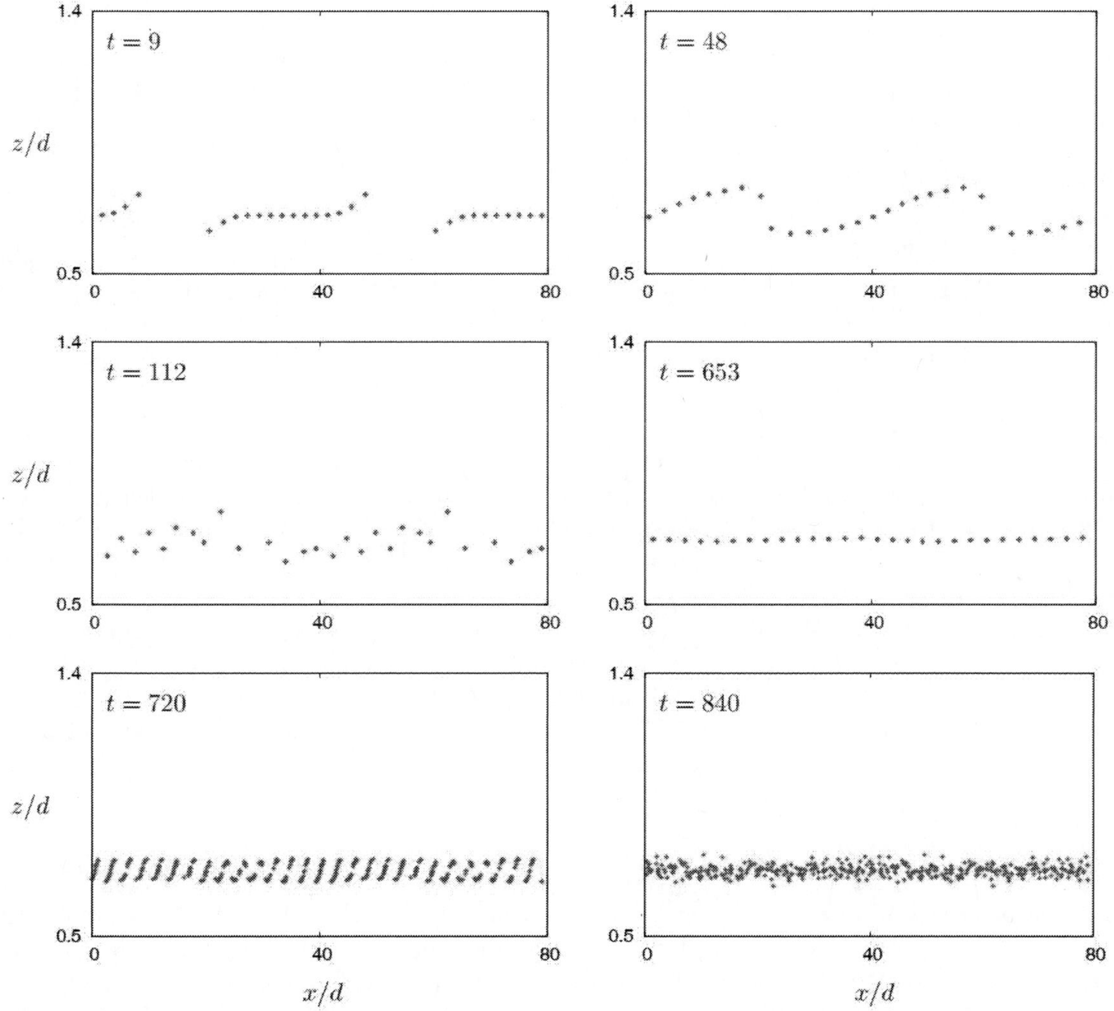

Figure 6. Evolution of a particle monolayer (side view). The panels show the projection of the particle centers onto the x–z plane at times as labeled. Note that the vertical scale is magnified. (Also see supplemental video movie 3.mov.)

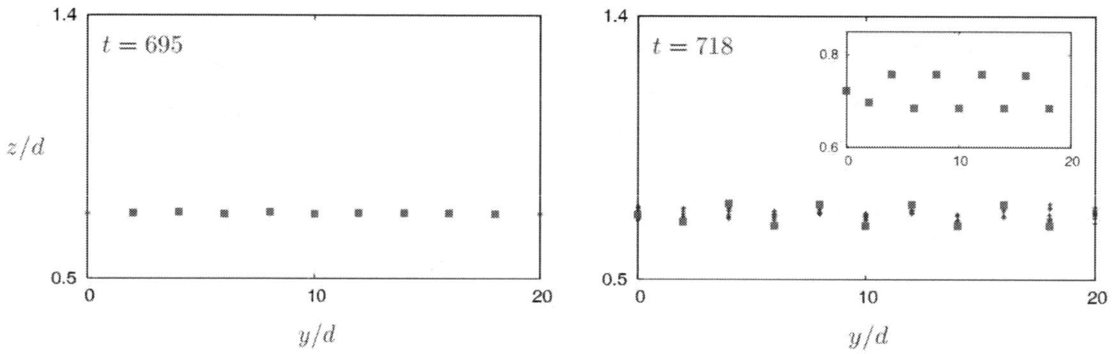

Figure 7. The buckling instability event (front view). The panels show the projection of particle centers onto the y–z plane at times as labeled. Particles in the front row are represented by red squares and all other particles by blue crosses. The inset shows positions of the front-row particles on a magnified scale. (Also see supplemental video movie_4.mov.)

can be easily observed in the side view (i.e., the flow–gradient x–z plane) of the evolving system. In Fig. 6 we show positions of particle centers in the x–z plane at the times corresponding to the frames depicted in Fig. 2. We note that for configurations where particle rows are perfectly aligned in the y direction (which is the case for $t = 9$, 48, 112, 653) only the centers of the front particles are visible in the figure.

The results shown in Fig. 6 for $t = 9$ are consistent with the predictions of our analysis of the ST mechanism (illustrated in Fig. 5). The front end of the strips moves upwards and speeds up, and the trailing end moves down and slows down. After the front row of particles in a given strip catches up with the trailing row of the next strip (cf. frame $t = 48$), significant ST interactions are established between these rows, and therefore, the entire rows change the direction of their vertical motion.

Subsequently, mutual ST interactions between neighboring particle rows give rise to their random motion in the vertical direction z and lead to corresponding fluctuations of the row distances in the flow direction x (cf. frames $t = 112$ in Figs. 2 and 6). The incoherent vertical motion of particle rows causes gradual cancellation of the fluctuations, and this cancellation results in a well-aligned rectangular lattice with regular interparticle spacing (cf. frames $t = 653$ in Figs. 2 and 6).

However, the lattice instability that takes place at $t = t_b \approx 700$ (cf. Fig. 3) entirely changes the system behavior: not only does a transition from a regular crystalline-like suspension microstructure to a disordered state occur, but also the particle alignment in the y direction suddenly breaks down (cf. frames $t = 720$ and $t = 840$ in Figs. 2 and 6). We examine the nature of this buckling instability in Sec. 2.3.

2.3. Buckling instability

Figure 7 shows the front view of the system just before the lattice instability event ($t = 695$) and just after the onset of the lattice breakup process ($t = 718$). Before the instability occurs the particles are well aligned along lines parallel to the flow direction x. The lack of vertical displacements indicates that the fluctuations observed at the earlier stages of the evolution have dissipated nearly completely. However, after the instability event significant vertical displacements take place (although the particles still preserve some alignment in the direction x).

In the new transient microstructure of the particle monolayer every other x-row of particles

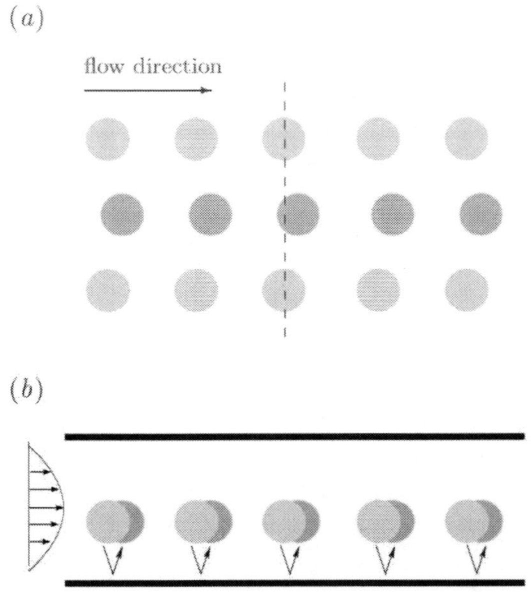

Figure 8. Buckling instability mechanism: a particle row that is ahead due to a random displacement (dark particles) receives a lift produced by the flow scattered from the particles in the neighboring rows that stay behind (light). The dark row is thus pushed up into a faster moving streamline and gets ahead even more. (a) Top view of the system, i.e., in the plane x–y; (B) side view.

is displaced up and every other x-row is displaced down, as seen in the right panel of Fig. 7. A comparison of the frames at $t = 695$, 710, and 720 in Fig. 3 indicates that x-rows of particles that have been displaced up move faster than the others. The consecutive x-rows of particles are displaced in opposite directions, and this buckling instability causes the decomposition of the previously well-aligned particle monolayer.

Qualitatively, the hydrodynamic mechanism of the buckling instability can be explained as follows. If a given x-row of particles is displaced up (i.e., away from the closer wall), it starts to move faster, and gets slightly ahead of the other x-rows, as illustrated in Fig. 8(a). Particles in the neighboring rows that are slightly behind produce scattered flow that gives lift to the upper displaced row of particles (similar to the lift that causes the ST effect depicted in Fig. 5). As a result, the displaced x-row moves even further from the wall, and speeds up more, leading to a rapid growth of the initial perturbation.

We note that a displacement of an individual particle does not produce a similar effect. The displaced faster moving particle approaches the particle in front of it and slows down, due to the ST interaction. Therefore, a one-particle perturbation does not grow. The unstable modes through which the particle lattice decomposes thus involve coherent displacements of entire segments of x-rows of particles, consistent with the transient structure of the system seen in Fig. 2 at $t = 720$.

An important issue that remains unsolved is the mechanism through which the particle lattice is stabilized against buckling at the earlier stages of the evolution. Since the only difference between the unstable state with well-aligned particles at $t = t_b$ and the fluctuating stable states for $t < t_b$ is the existence of lateral and transverse displacements of y-rows of particles, such displacements must be responsible for the array stabilization. The behavior of the particle monolayer qualitatively resembles stabilization of an elastic sheet by a transverse deformation wave. Such a wave, oriented along one direction, prevents buckling of the sheet in the orthogonal direction. While this analogy provides some useful intuition, a specific hydrodynamic mechanism stabilizing the particle array still needs to be investigated.

3. Conclusions

Our numerical simulations of an initially ordered particle monolayer driven by Poiseuille flow in a parallel-wall channel have revealed very interesting and rich dynamics. As depicted in Figs. 2 and 6 the system can heal large gaps in the particle lattice, and lattice fluctuations can spontaneously decay, leading to a well-aligned rectangular configuration. We have also described a new buckling instability of the particle monolayer, and prior lattice stabilization through displacements of particle rows orthogonal to the buckling direction. All of these phenomena, which occur in the Stokes-flow regime, are purely hydrodynamic in their nature and arise from the ST effect.

Our simulations have demonstrated the existence of a new hydrodynamically stabilized disordered suspension phase. In this phase, particles are arranged into meandering strings, with orientations changing between the flow direction x and the vorticity direction y. This phase (briefly discussed in our earlier study [28]) closely resembles the microstructure of particle layers that spontaneously form in oscillatory shear flow, as observed in recent experiments [27]. These experiments also show spontaneous formation of a *square* particle lattice (as opposed to a hexagonal configuration resulting from excluded-volume effects). The structures in our simulations and in experiments [27] are strikingly similar, although the respective results were obtained for different flows and system geometries. Therefore, the observed patterns (i.e., hydrodynamically stabilized square lattice and string-like particle correlations) underscore the generality of our findings.

Our analysis of the dynamics of a particle monolayer has revealed an important role of the ST effect. The ST mechanism, which influences the system at all stages of its evolution, is associated with wall-mediated hydrodynamic interactions between particles in a local shear flow. The ST phenomenon causes cross-streamline displacements of approaching particles: the particles move in opposite directions normal to the wall, and as a result of this position swapping across the streamlines, the particles reverse their relative motion and separate without collision.

The multiparticle ST interactions in confined suspension flows have far-reaching and diverse consequences. On the one hand, the ST mechanism stabilizes fluctuating quasi-2D monolayers by preventing particle collisions. On the other hand, it has a destabilizing effect in perfectly ordered 2D monolayers by causing buckling of the particle lattice. The resulting string-like microstructure is, again, stabilized by the ST effect. Other phenomena produced by the ST effect were discussed in our earlier studies: the ST mechanism can produce spontaneous suspension layering [24], and it can lead to enhanced hydrodynamic diffusion [21].

While our study is focused on the dynamics of infinite quasi-2D particle monolayers, we expect that similar phenomena also occur in microfluidic flows that involve particle arrays of finite extent. In particular, we predict stabilization and realignment of linear particle trains via cancellation of random fluctuations, i.e., due to a mechanism similar to the one seen in Fig. 6 for $t \leq 653$. (In fact, this mechanism is likely to play a significant role in the recently observed spontaneous particle ordering at non-zero Reynolds numbers [14, 15].) We propose that experiments to observe the buckling effect (illustrated in Fig. 7) can be carried out using a multi-orifice flow-focusing devices [36] to generate synchronized parallel particle trains in a channel with a rectangular crosssection of high aspect ratio. Such experiments would help to explore the diverse consequences of the ST effect in confined particulate flows.

Acknowledgments

3.1. Acknowledgments

We acknowledge financial support by NSF grant CBET 1059745 (JB) and Polish Ministry of Science and Higher Education grant N N501 156538 (EW). JB wishes to thank François

Feuillebois for exposing him to the science of wall-bounded suspensions and to the art of French living during the author's stay in Laboratoire d'Aérothermique du CNRS and ESPCI in 1993 and 1994.

References

[1] Riedel I H, Kruse K and Howard J 2005 *Science* **309** 300–303
[2] Lenz P and Ryskin A 2006 *Phys. Biol.* **3** 285–294
[3] Underhill P T, Hernandez-Ortiz J P and Graham M D 2008 *Phys. Rev. Lett.* **100**
[4] Garstecki P and Whitesides G M 2006 *Phys. Rev. Lett.* **97** 024503
[5] Hashimoto M, Mayers B, Garstecki P and Whitesides G M 2006 *Small* **2** 1292–1298
[6] Beatus T, Tlusty T and Bar-Ziv R 2006 *Nature Phys.* **2** 743–748
[7] Cui B, Diamant H, Lin B and Rice S A 2004 *Phys. Rev. Lett.* **92** 258301–1–4
[8] Koppl M, Henseler P, Erbe A, Nielaba P and Leiderer P 2006 *Phys. Rev. Lett.* **97** 208302
[9] Garstecki P 2006 *Nature Phys.* **2** 733–734
[10] Beatus T, Tlusty T and Bar-Ziv R 2007 *Phys. Rev. Lett.* **99** 124502
[11] Hashimoto M, Garstecki P, Stone H A and Whitesides G M 2008 *Soft Matter* **4** 1403–1413
[12] Baron M, Bławzdziewicz J and Wajnryb E 2008 *Phys. Rev. Lett.* **100** 174502
[13] Bonthuis D J, Meyer C, Stein D and Dekker C 2008 *Phys. Rev. Lett.* **101** 108303
[14] Humphry K J, Kulkarni P M, Weitz D A, Morris J F and Stone H A 2010 *Phys. Fluids* **22**
[15] Lee W, Amini H, Stone H A and Di Carlo D 2010 *PNAS* **107** 22413–22418
[16] Pasol L, Martin M, Ekiel-Jezewska M L, Wajnryb E, Blawzdziewicz J and Feuillebois F 2011 *Chem. Eng. Sci.* **66** 4078–4089
[17] Bruneau D, Feuillebois F, Anthore R and Hinch E J 1996 *Phys. Fluids* **8** 2236–2238
[18] Bławzdziewicz J and Feuillebois F 1995 *Fluid Mech. Research* **22** 66
[19] Bruneau D, Feuillebois F, Bławzdziewicz J and Anthore R 1998 *Phys. Fluids* **10** 55–59
[20] Zarraga I E and Leighton D T 2002 *Phys. Fluids* **14** 2194–2201
[21] Zurita-Gotor M, Bławzdziewicz J and Wajnryb E 2007 *J. Fluid Mech.* **592** 447–469
[22] Alvarez A, Clement E and Soto R 2006 *Phys. Fluids* **18** 083301
[23] Mylyk A, Meile W, Brenn G and Ekiel-Jezewska M L 2011 *Phys. Fluids* **23**
[24] Zurita-Gotor M, Bławzdziewicz J and Wajnryb E 2012 *Phys. Rev. Lett.* **108** 068301
[25] Vananroye A, Van Puyvelde P and Moldenaers P 2006 *Langmuir* **22** 2273–2280
[26] Tufano C, Peters G W A and Meijer H E H 2008 *Langmuir* **24** 4494–4505
[27] Cheng X, Xu X, Rice S A, Dinner A R and Itai C 2012 *PNAS* **109** 63–69
[28] Blawzdziewicz J, Goodman R H, Khurana N, Wajnryb E and Young Y N 2010 *Physica D* **239** 1214
[29] Bhattacharya S, Bławzdziewicz J and Wajnryb E 2005 *Physica A* **356** 294–340
[30] Cichocki B, Felderhof B U, Hinsen K, Wajnryb E and Bławzdziewicz J 1994 *J. Chem. Phys.* **100** 3780–3790
[31] Bhattacharya S, Bławzdziewicz J and Wajnryb E 2006 *J. Comput. Phys.* **212** 718–738
[32] Bławzdziewicz J and Wajnryb E 2008 *Phys. Fluids.* **20** 093303
[33] Bhattacharya S, Bławzdziewicz J and Wajnryb E 2005 *J. Fluid Mech.* **541** 263–292
[34] Bossis G, Meunier A and Sherwood J D 1991 *Phys. Fluids A* **3** 1853–58
[35] Janssen P J A, Baron M D, Anderson P D, Blawzdziewicz J, Loewenberg M and Wajnryb E 2012 Collective Dynamics of Confined Rigid Spheres and Deformable Drops arXiv:xxx
[36] Hashimoto M, Shevkoplyas S S, Zasonska B, Szymborski T, Garstecki P and Whitesides G M 2008 *Small* **4** 1795–1805

Journal of Physics: Conference Series **392** (2012) 012009 doi:10.1088/1742-6596/392/1/012009

Hydrodynamics in nanoscale confinement: SFA and colloid probe AFM liquid drainage experiments

M. Krasowska[1], M. N. Popescu, and J. Ralston

Ian Wark Research Institute, University of South Australia, Adelaide, SA 5095, Australia

E-mail: Marta.Krasowska@unisa.edu.au, Mihail.Popescu@unisa.edu.au, John.Ralston@unisa.edu.au

Abstract. Flow and drainage of very thin liquid films play an important role in mineral recovery, drop coalescence and emulsion stability, as well as lubrication of micromechanical devices. Studies of liquid flow under strong confinement (i.e., film thickness below a few hundred of nanometers and down to a few nanometers) can reveal the limits of applicability of a classical hydrodynamics description, but are very challenging. The Surface Force Apparatus (SFA) technique has enabled studies of drainage at nanoscale separation between atomically smooth mica sheets. The development of the colloid probe Atomic Force Microscope (AFM) as an alternative technique has allowed a significantly wider variety of confining solid surfaces to be studied. Both the SFA and the colloid probe AFM have been adapted to permit the surfaces confining the film to be soft, e.g., the surface of a drop or bubble, and therefore deformable. We present a succinct review of the experimental and theoretical modeling challenges for such studies and critically discuss the outcomes of recent experiments.

Dedicated to Prof. François Feuillebois on the occasion of his 65th anniversary

1. Introduction

Flow and drainage of very thin liquid films play an important role in, among others, mineral recovery by flotation, drop coalescence and emulsion stability, as well as lubrication of micromechanical devices. The rate of drainage of a Newtonian liquid in situations where the confinement is strong, i.e., when the surfaces bounding and squeezing out the film are separated by distances below a few hundred of nanometers and down to few nanometers, is also of interest for understanding the limits of applicability of classical hydrodynamics. Moreover, such studies can eventually provide a way to assess the validity of the hydrodynamic boundary conditions, for example the no-slip condition usually assumed for liquid flow over a smooth solid surface. A violation of the no-slip boundary condition can be quantified in terms of the so called hydrodynamic slip length λ (see, e.g., Ref. [1]). Slip can have an observable effect on the drainage rate only when the lengthscale of confinement becomes comparable with λ. For smooth surfaces (roughness smaller than few nm), it has been argued that the slip length, if any, should range from no more than a few nm for lyophilic surfaces to a maximum of few tens of nm for lyophobic ones [1,2]. Sophisticated techniques and very well controlled experimental conditions are therefore necessary to ensure the required strong confinement and to enable a reliable discrimination of a slip length.

[1] Corresponding authors

Various methods, such as bubble-plate collision [3–6], the Thin Film Balance (TFB) balance technique (also known as the Sheludko cell) [7, 8], the Surface Force Apparatus (SFA) [9–13], and the colloid probe Atomic Force Microscope (AFM) [14–25] have been employed to study hydrodynamic behaviour in confined systems. In this paper we will focus on investigations with Newtonian liquids involving the latter two methods. [2]

1.1. *Thin film drainage in SFA and colloid probe AFM*

A large number of studies used the SFA to study drainage of various liquids between atomically smooth mica sheets (see, e.g., Refs. [9, 13, 26, 27]. Almost two decades ago the colloid probe AFM technique was developed [14]. This permitted the extension of thin film drainage studies to a much larger variety of solids (see, e.g., Refs. [17, 18, 20, 21, 28, 29] and references therein), including the so-called asymmetric systems, i.e., the two bounding solid surfaces being made out of different materials, e.g., mica and silica, silica and titania, or silica and polystyrene. A consistent theoretical framework for the interpretation of these thin film drainage experiments was developed in Ref. [30]. At about the same time, modifications of the SFA have been made to allow one of the surfaces to be soft and thus deformable, e.g., a drop or a bubble [31, 32]. Since then, a number of studies involving such set-ups have been carried out (see, e.g., Ref. [33] and references therein). In the last decade the colloid probe AFM has been also extended to permit studies in which one or even both of the surfaces confining the film are soft (see, e.g., Refs [22–25, 34–36] and references therein). Building upon the theoretical background developed in Refs. [31, 37–39] to describe the static surface forces in the case when one of the two interfaces is soft, in the last decade a very elaborate theoretical framework aiming at a consistent interpretation of the AFM measurements of film drainage between deformable surfaces was developed (see Refs. [40–42] for an excellent, detailed presentation).

In Fig. 1 we show schematically the typical experimental setup of the AFM or SFA thin film drainage studies. We note that the relative velocities of the two bodies in the typical AFM or SFA experiment are rather small (usually not greater than few tens of micrometers per second). The corresponding Reynolds number $\mathrm{Re} = \rho_f U h_0/\mu_f$ (where ρ_f and μ_f are the density and viscosity of the liquid, respectively, U is the average flow velocity, and $h_0 := h(0,t)$ is the separation) is small; for example, in the case of a water film draining with an average flow velocity $U \simeq 1$ mm/s (corresponding to surfaces approaching at speeds of the order of micrometers/s) at separation $h_0 = 10$ nm, the Reynolds number is $\mathrm{Re} \simeq 10^{-5}$. The hydrodynamics of a Newtonian fluid is then well-described by the Stokes equations [17, 20, 42, 43]. (We note here that the Newtonian character of the liquid under such strong confinement is actually a model assumption which should be validated independently.) Moreover, in the range of film thicknesses of interest here, i.e., $h \lesssim 100$ nm the lateral extent of the interaction zone, which is defined as the part of the system where the pressure differs significantly from that of the bulk liquid [42], is of the order of microns or more (see Fig. 1), i.e., much larger than the film thickness. This allows one to describe the film drainage within the approximations of the Reynolds lubrication theory.

The corresponding thin-film hydrodynamics equation, basically employed for the data analysis by all the AFM and SFA studies that we discuss here, is written in cylindrical coordinates $\{r, \phi, z\}$ (see Fig. 1) as [30, 41, 42]

$$\mu_f \frac{\partial^2 u}{\partial z^2} = \frac{\partial p}{\partial r}, \tag{1}$$

where $p := p(r,t)$ denotes the pressure field and $u := u(r,z,t)$ is the flow field. This is to be

[2] The use of SFA and colloid probe AFM techniques in studies of hydrodynamics under nanoscale confinement was one of the points of "Challenges in microhydrodynamics" session of the workshop "Microparticles in Stokes flows", August 21-24 2011, Warsaw (Poland).

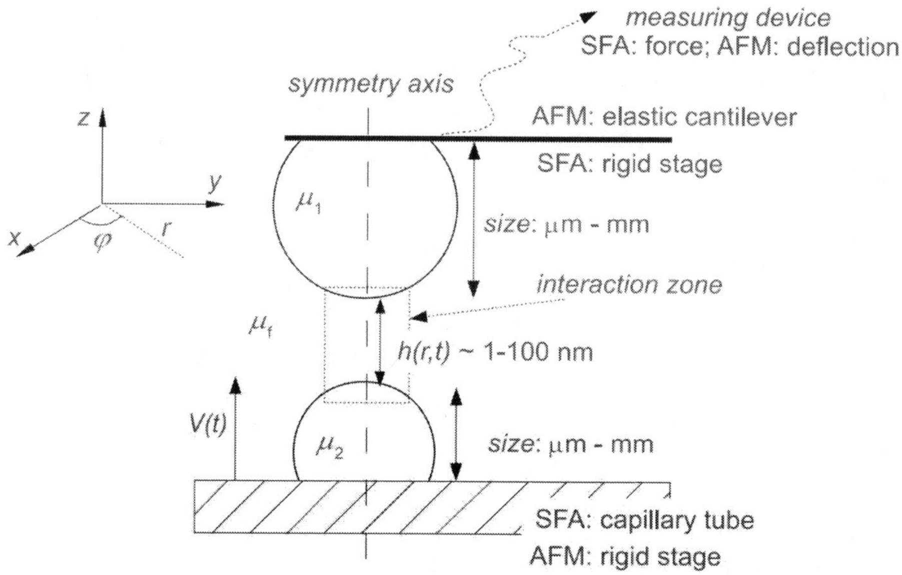

Figure 1. Schematic drawing of a typical AFM or SFA liquid drainage experiment. The two bodies of viscosities $\mu_{1,2}$ are moved one towards another through a fluid of viscosity μ_f by displacing either the lower stage (as shown here), or the top one, with an externally prescribed velocity $V(t)$. Either or both of the two bodies can be spheres, part of spheres, or flats (infinite radius of curvature). In the modified SFA discussed in Refs. [31,33] the lower stage is replaced by a rigidly fixed capillary filled with the fluid 2, which is kept at a constant pressure that is sufficient for creating the spherical protrusion, as shown in the figure. The case of a solid body is recovered by taking the limit $\mu_i \to \infty$ ($i = 1$ or 2), while that of fluid corresponds to finite values of the viscosity. The values $h(r,t)$ shown in the figure indicate the range of interest for the present work. The top stage or cantilever is connected to a device that measures the force $F(t)$, either directly (SFA) or by converting the measured cantilever deflection (AFM), experienced by the body 1 during the motion. The experimental setup involves a careful alignment such that the collision is "head-on" and the system has rotational symmetry around the axis indicated in the figure. The distance r is the polar radius measured from this axis. (For clarity, the coordinates system is shown shifted to the left of the symmetry axis, rather than with the z-axis coinciding with it.)

solved within a domain $0 \le r \le r_{max}$ subject to appropriate boundary conditions on the surfaces $z_{1,2}(r,t)$ of the two bodies, which define the film thickness as $h(r,t) := z_1(r,t) - z_2(r,t)$.

If the two surfaces are rigid, the lateral extent r_{max} is determined from matching the solution for the flow inside the thin film with the macroscopic hydrodynamic flow around the two bodies [30]. (The case of small elastic deformations of the two rigid bodies has also been discussed in Refs. [19,20].) If one or both surfaces are soft, an additional condition, that of mechanical equilibrium for the shape of the *fixed volume* deformable body (the so called "augmented Young-Laplace equation" [41,42]), is necessary in order to determine the value of r_{max}. The analysis, in particular in the cases of deformable bodies, is quite involved. For the purpose of our paper the details of these derivations are not needed; therefore we will not review them here, but we encourage the interested reader to consult, e.g., the Refs. [19,20,30,41,42].

Under the assumption of incompressibility, the thickness $h := h(r,t)$ of the film obeys the equation [30, 41, 42]

$$\frac{\partial h}{\partial t} = -\frac{1}{r}\frac{\partial}{\partial r}\left(r \int\limits_0^{h(r,t)} u\, dz \right) . \tag{2}$$

For two rigid bodies, this couples directly with the externally imposed motion of the stage or that of the cantilever and with the cantilever deflection D as $\frac{\partial h}{\partial t} = -V(t) + \frac{dD}{dt}$ (positive deflection of cantilever is chosen to mean above its equilibrium position). For the case in which at least one of the bodies is deformable, the external motion and the cantilever deflection enter the description via a coupling condition for the thickness $h(r_{max})$ at the edge of the interaction zone [41, 42]. The total force acting on the body 1 and measured by the force measuring device is determined as [42]

$$F(t) = 2\pi \int\limits_0^{\infty} [p(\xi, t) + \Pi(\xi, t)]\, \xi\, d\xi\,, \tag{3}$$

where $\Pi(\xi, t) := \Pi(h(\xi, t))$ denotes the so-called disjoining pressure (the negative of the derivative, in respect to the separation, of the excess free-energy of two flat interfaces being in close proximity to one another and the polar radius ξ is measured from the symmetry axis of the system. Note that Eq. (3) implicitly uses the proximity approximation in that the total contribution of the disjoining pressure is written as a linear superposition of contributions arising from the parts of film at each position ξ with the corresponding thicknesses $h(\xi, t)$. Therefore once (i) the viscosity μ_f of the intervening liquid, (ii) the hydrodynamic boundary conditions on the surfaces 1 and 2, and (iii) the expression of the disjoining pressure $\Pi(h, t)$ are given, the force $F(t)$ and the films thickness $h(r, t)$ in the interaction zone can be computed. The experimentally accessible quantities are $F(t)$ and, in the SFA measurements, $h(r, t)$, as functions of time or, as typically done for the AFM experiments, of the stage displacement $\Delta X(t) = \int_0^t dt'\, V(t')$.

The experimental studies mentioned in the beginning of this section, together with the development of a theoretical background for the interpretation of the results, have significantly contributed to a better understanding of thin film drainage under nanoscale confinement. However, a number of aspects are yet to be clarified. In Sec. 2 we briefly review the theoretical modeling and experimental challenges faced by the thin film drainage studies with colloid probe AFM or SFA. We then critically discuss in Sec. 3 the outcomes of the recent experiments, including the issue of an unexpected no-slip boundary condition at a fluid-fluid interface which seems to be needed for a quantitative description when soft, deformable interfaces are involved. Finally, our conclusions and a brief outlook at future developments are presented in Sec. 4

2. Intrinsic theoretical and experimental challenges.

When a comparison of the theoretical predictions for the time dependence of the force $F(t)$ or of the shape $h(r, t)$ of the film with the experimental data is attempted, a number of challenging issues for both the theory and the experiment occur. We discuss below those that we consider most relevant.

2.1. Theoretical challenges

In Eq. (1) the viscosity μ_f of the liquid film influences the time scale of the film thinning (drainage rate). Once the film thickness is in the nanometer range, i.e., $h(r, t) \lesssim 100$ nm, it is no longer *a priori* clear that the viscosity is the same as that of the bulk liquid, nor even that the assumption of a Newtonian liquid behavior remains valid. For molecularly thin films a stick-slip behaviour and shear-thinning or shear-thickening transitions are well documented in the literature (see, e.g., Refs. [9, 13, 26, 27]). The analysis of the AFM and SFA thin films experiments thus must be complemented by independent measurements of the viscosity of the liquid in thin films, if the aim is to probe the validity of Stokes equations under nanoscale confinement or specific hydrodynamic boundary conditions.

The dependence of the total force on the disjoining pressure $\Pi(h)$, Eq. (3), requires that the various contributions, e.g., the electrostatic double layer and the van der Waals classic DLVO interactions, are known precisely in order for a quantitative comparison between theory and experiment to be possible. A quantitative calculation of the classic DLVO interactions remains a challenging point because: **(i)** the exact phenomenology of the charging of the interface, which determines the boundary conditions for the Poisson-Boltzmann equation, is generally unknown and at best accounted for by a so-called "charge regulation" boundary condition dependent on parameters to be determined by fitting experimental data (see Ref. [44] and references therein). For soft interfaces, for example that between a bubble and a salt solution, reliable values of these parameters are notoriously difficult to find in the literature (see, e.g., the discussion in Ref. [25]); **(ii)** the van der Waals interactions require precise knowledge, and are very sensitive to the details, of the dielectric permittivity functions of the two bodies and of the liquid. For water, for example, precise information is not yet available and various constructions of the dielectric permittivity function, giving rise to significantly different interactions, are *a priori* plausible (see Ref. [45]). Even for the most well-studied systems, those involving mica surfaces, the agreement between theoretical predictions and the experimental measurements of van der Waals interactions is seldom better than within 30% [46]; **(iii)** in the case when the liquid is a salt solution, which is the case in many of the experimental studies, [3] it has been argued that a screening effect can occur even for the van der Waals interactions [52]. The exact form of the screening is presently under investigation, while only a few experimental studies of such effects have been reported, with so far rather inconclusive results [45, 46, 53]; **(iv)** a theoretical description of dynamic surface interactions, i.e., the time dependent equivalent of the DLVO theory, is yet to be developed. The validity of the quasi-static approximation, i.e., using for the disjoining pressure the equilibrium expression at the instantaneous film thickness $h(r,t)$, which is invoked in order to close the Eqs. (1)-(3), can be assessed only *a posteriori*. This is even more stringent in cases when the liquid separating the two bodies is actually a salt solution, because the convective transport of the ions in films of small thickness may disrupt the double layer charge distributions (which relax diffusively).

The hydrodynamic flow field $u(r,t)$ in Eq. (1) is determined subject to boundary conditions on the two surfaces. For a liquid-solid interface, the condition on the tangential velocity is *postulated*: either a no-slip boundary condition, i.e., $u \equiv 0$ at the surface of the solid, which is the generally accepted one for macroscopic hydrodynamics, or a Navier slip boundary condition, i.e., $u = \lambda du/dn$ at the surface of the solid, where n denotes the direction normal to the surface of solid and oriented into the liquid, while λ is the slip length (note that in the limit $\lambda \to 0$ the no-slip boundary condition is recovered). There exists no general prescription of how to calculate the slip length λ for a given liquid-solid pair [54], and therefore λ remains a fitting parameter for all the models of film drainage in AFM and SFA experiments. For a fluid-fluid interface, the hydrodynamic boundary conditions, also known as "mobile interface" boundary conditions, are the continuity of the tangential velocity and of the tangential shear stress (as well as continuity

[3] Salt solutions are extensively used in experimental studies because of the possibility to control the range of the electrostatic interactions via the salt concentration. Unfortunately, they are also subject to potentially significant complications such as, e.g., the fact that the multivalent and the asymmetric electrolytes are poorly described by the mean-field Poison-Boltzmann theory (see, e.g., Ref. [47] and the references therein). Moreover, the ion-specificity, i.e., the so-called "Hofmeister effect" (see, e.g., Ref. [48] and references therein) has been a puzzle for almost a century, and it is only recently that a consistent theoretical description for it seems to emerge [49–51].

of the normal velocity and the Laplace jump in pressure due to surface tension and curvature of the interface); they are *determined* (rather than being postulated) by the requirement that such interfaces cannot sustain any tangential stress. If a third species (e.g., surfactants) is present at a fluid-fluid interface, the interface can become "partially" or "completely" immobilized, i.e., the boundary conditions may change into one of partial or no slip [42]. Even in the absence of such surface active contamination an effective no-slip boundary condition may also apparently hold if the ratio of the viscosities of the two fluids satisfies certain conditions [42]. Therefore the choice of the hydrodynamic boundary conditions remains in most cases a parameter of the theoretical modeling.

2.2. Experimental challenges

The experimental challenges occurring when employing the colloid probe AFM technique in drainage studies have been thoroughly reviewed in the recent Ref. [17, 28], as well as in Refs. [29, 55]. Here we discuss only few of the experimental issues occurring in a typical thin-film drainage experiment.

Whenever solid surfaces are employed in the SFA or AFM studies, and in particular the spherical probes in the AFM, the issue of the surface roughness typically cannot be avoided. Except for the case of the mica sheets, which are atomically smooth (and also sufficiently elastic to be glued to SFA quartz lenses), for most of the other materials used as planar substrates or as colloid probes the surface is rough: (root-mean-squared) roughness of at least 0.5 nm and peak-to-valley values of at least a couple of nanometers. This has several implications, all of which make very difficult (if at all meaningful) a quantitative comparison between the experimental results and the theoretical predictions obtained under the assumption of smooth, flat surfaces [18,56]. **(i)** Contact between such surfaces, thus the zero-separation point required for the linear-compliance calibration of an AFM, can be defined only in a statistical sense. **(ii)** Only an "effective" separation $\bar{h}(r,t)$, measured between the average (in respect to roughness) shape of the surfaces can be defined. Because the main interest is in film drainage at (effective) separations in the nanoscale range, e.g. ca 10 nm and below, even 1-2 nm peak to valley roughness can lead to significant (10% - 20%) differences between the separation in the experiment and the "effective" one employed in the theoretical modeling, both in Eq. (2) and in the definition of disjoining pressure. **(iii)** In the presence of rough surfaces it is not clear where the "effective" surface lies and what are the hydrodynamic boundary conditions which should apply there. The commonly taken approach is to apply a slip boundary condition at the position of the "average" surface [1,18]; obviously, the slip-length occurring in such a boundary condition remains a parameter which is very difficult to connect with the microscopic details (peak-to-valley values, lateral correlation length of roughness features, etc.). [4] Moreover, when dissolved gas is present in the fluid (and this is the case of most of the experimental situations), nanoscale bubbles may nucleate in the valleys of the rough topography, giving rise to a heterogeneous surface partially covered by bubbles. In this case, Ref. [65] has shown that the slip length can be either positive, which intuitively means a smaller resistance to liquid drainage than that for a no-slip surface, or become negative, which would mean a larger resistance than that of a no-slip

[4] We note here that recently a number of theoretical and numerical studies investigating the "effective" slip-length for model surfaces with well defined distributions of patches with different slip-lengths (usually two-values distributions) have been published. Significant progress has been made in this direction, including: calculations of the effective slip-lengths for flat surfaces with periodically distributed "stripe-like" patterns of given slip-lengths [54, 57], predictions and exact calculations of a tensorial slip boundary condition for anisotropic surfaces (e.g., containing topographical grooves, or stripes of varying slip-lengths) [58], exact bounds for the effective slip lengths for flows in thin-channels with textured walls [59], theoretical calculations and Lattice-Boltzmann simulations results for the effective slip lengths on model superhydrophobic surfaces [60–62] as well as for the drainage rate between a smooth sphere and model anisotropic or superhydrophobic flat (on average) surfaces [63, 64].

surface. **(iv)** A rigorous calculation of the DLVO surface interactions between rough surfaces when the separation between the surfaces is of the order of the peak-to-valley roughness is not available yet.

As mentioned in Sec. 1, in the colloid probe AFM technique only the deflection of the cantilever is measurable, through a deflection of a reflected laser beam, from which the force is calculated based on a calibration of the elastic constant of the cantilever and of the voltage response of the photo-detectors. At small separations between the surfaces the forces become large (and have steep spatial gradients), and the accuracy of the measurement, which depends on the assumption of a linear response of the photo-detector, should be carefully considered. The deflection of the cantilever can be eventually converted into the central separation $h(0, t)$ between (smooth) *rigid* surfaces based on the so-called "linear compliance regime". This requires the surfaces to be brought into real mechanical contact, and therefore can be applied only when the two surfaces are rigid. But for rough surfaces this procedure cannot lead to more than just an estimate, rather than an absolute value, for an effective separation; obviously, this is particularly important at very small effective separations (e.g., 10 nm or less). For soft, deformable surfaces, no information about the separation $h(r, t)$ can be obtained with the present set-up of the colloid probe AFM. In contrast, the SFA apparatus allows for a direct (and independent) measurement of the separation $h(r, t)$ between the two surfaces even when they are deformable by using fringes of equal chromatic order (FECO) interferometry. However, for solid surfaces which are rough the challenges listed for the AFM techniques are also relevant; additionally, the interpretation of the FECO interferometry spectra is no longer straightforward.

For colloid probe AFM, the motion of the cantilever through the liquid is subject to viscous friction and gives rise to an additional deflection. In general, this cannot be rigorously calculated because the shape and the material properties of the cantilever are too complex. Moreover, the interplay between length-scales: micrometer, for the length and thickness of cantilever and the diameter of the body 1, and nanometer for the film thickness separating the two bodies, makes a numerical calculation of the viscous friction on the cantilever a very difficult task.

Finally, contamination is an issue for both SFA and AFM experiments. For fluid-fluid interfaces, traces of surface active contaminants are sufficient to change the fully mobile character of the interface into a (partially) rigid-interface one (see Ref. [42] and references therein). For solid-liquid interfaces, insoluble nano-particles (e.g., dust) can adsorb on the solid surface; this gives rise to a spurious smaller value of the "measured" film thickness because the mechanical contact occurs on the contaminant particle rather than on the surface, a feature which by now is very well documented in the literature [17, 28, 66, 67]. The presence of dissolved gases and nucleation of nano-bubbles not only affects the hydrodynamic boundary conditions, but can also give rise to a van der Waals attraction between bubbles on the opposite surfaces [17, 28]. Such an effect is an example of the so-called "long-ranged hydrophobic attraction". This affects the value of the disjoining pressure in Eq. (3), but its occurrence cannot be *a priori* assessed and the magnitude of the effects cannot be exactly quantified.

3. Investigations of thin film drainage using the SFA and the colloid probe AFM

In view of the various challenges and issues briefly reviewed in the previous section we summarize here the results of the SFA and AFM studies in the last two decades, with a focus on their interpretation as a test for the hydrodynamic description of film drainage.

The early colloid probe AFM experiments employed sucrose solutions, silica probes and muscovite mica surfaces covered with layers of titania and gold, on top of which self-assembled monolayers of certain chemical reagents have been used to tune the "hydrophobicity" of the surfaces [16, 68, 69]. The strategy followed was to use high velocities of the AFM stage such that the hydrodynamic effects are significant and eventually dominant over the disjoining pressure contributions [see Eq. (3)]. The authors claimed to observe good agreement with the theoretical predictions reported in Refs. [30] but this is achieved *by fitting the experimental data with slip lengths that are dependent on the shear rate* (presumed to be constant in an experiment at a given nominal approach speed of the AFM stage). This is unfortunately inconsistent with the theoretical model, a point clearly made in Refs. [15, 21, 29, 42]. Moreover, these experiments have been seriously questioned for a number of possible problems with the experimental set-up and artifacts. Problems include those of too soft cantilevers and contamination of surfaces by nano-particles, as identified and discussed by Refs. [15, 20, 21, 29, 70]; these also report for the same systems, as well as for other liquid and solids, and for similar dynamic conditions (fast motion of the AFM stage) a no-slip, or very small slip lengths of up to few nanometers, boundary conditions. Recently, these pioneering experiments and the experimental set-up have been thoroughly reviewed [17, 28]; additional reports of no-slip boundary conditions at hydrophilic surfaces and of various degrees of slip at hydrophobic or rough surfaces have been made [17, 18, 28, 56]. In spite of the relatively large number of studies, the conclusions reached (slip or no slip) are often conflicting and the reported values for slip-lengths vary wildly between the various papers. Moreover, recently it has been observed that the shape of the cantilever actually influences the outcome of the experiment in terms of the type of boundary condition stemming out of the experimental data [55]; as correctly pointed out by Ref. [42], in the absence of a systematic study to understand the origin, and the exact magnitude, of the effect that the cantilever shape has on the outcome of the force measurements, this leaves doubt as to the reliability of all previous experiments.

The various SFA and colloid probe AFM experiments on thin film drainage between soft, deformable bounding surfaces, as well as experiments involving different experimental techniques, have been generally performed using approach velocities for which both the hydrodynamic pressure and the disjoining pressure contributions are relevant; these have been recently reviewed by Refs. [41, 42], with a view focused on the theoretical modeling. We therefore discuss here only few specific points and encourage the interested reader to consult these reviews and the extensive list of references included there. The most complete test of the theoretical models is provided by the SFA experiments, for both drainage and filling (i.e., bounding surfaces moving *towards* or *away from* each other, respectively) with drops and bubbles of Horn *et al* (see, e.g., Refs. [12, 31–33, 42] and the references therein) because they provide the whole thickness profile $h(r,t)$ as a function of time. According to Ref. [42], the theoretical modeling based on the Eqs. (1) - (3) and the augmented Young-Laplace equation (see Sec. 1) captured very well the qualitative features observed in the experiment, in particular: the occurrence of "dimple" and "wimple" shapes (i.e., regions of curvature with opposite signs) in the interaction zones, as well as the time scales and approach velocities over which these instabilities develop and relax, down to film thicknesses (minimum of $h(r,t)$) in the interaction zone in the order of 50 nm. In view of the complexity of the experimental results, these findings provide a strong argument that the hydrodynamic description reviewed in Refs. [41, 42] seems to capture the essential physics.

Furthermore, it has been shown that the same model can lead to predictions in quantitative agreement with the experimental results for: $h(r,t)$ in the SFA setup and the approach and retract force curves $\{\Delta X(t), F(t)\}$ in the AFM set-up. But this is achieved only if *a no-slip boundary condition* is invoked at the fluid-fluid interface(s). These findings hold for various configurations of the bodies 1 and 2: bubble-plate, drop-plate, drop-drop, or bubble-bubble, as well as for various choices of the intervening liquid (see, e.g., the review in Ref. [42]). [5] This is very surprising. For example, by using as an estimate the criteria expressed in Ref. [42] (p.

[5] We note, however, that Ref. [25] mentions that for quantitative agreement to be achieved for a few parameters slightly different values than the experimentally measured ones had to be used. It is unfortunately difficult to assess if this was also the case for other experiments because in many cases – see, e.g., Ref. [23] – no complete information is given about the values of the parameters, for example the spring constant of the cantilever, used in the theoretical modeling and experimentally measured.

2256) for an air bubble - water interface, the viscosities of the two fluids, the radius of the bubble in the experiment, and the film thickness in the range of tens of nanometers would argue in favor of a fully mobile interface. The most frequently invoked argument for such a no-slip boundary condition is that of surface active contaminants (see, e.g., Ref. [23]) which, even in minute amounts, would be sufficient to partially or fully immobilize the interface. However, this is unlikely to be the case in Refs. [12, 25, 31, 33], where explicit reference is made to significant efforts made to ensure the cleanness of the system; for example, for the bubble approaching a titania plate studied in Ref. [25] the presence of contaminants would likely make the titania surface hydrophobic and lead to the film rupture, which is not what is observed experimentally. The fact that the experimental conditions indeed can be controlled to ensure the absence of surface active contaminants is further supported by the reports of mobile boundary conditions for bubbles in water and aqueous electrolytes [6]. Moreover, a mobile boundary condition at a clean air-water interface has also been clearly evidenced independently in Ref. [71], which used a different experimental technique (the motion of optically trapped colloids near the air-water interface). The recent arguments that a no-slip boundary condition is needed to obtain quantitative agreement with the results of experimental bubble-rise studies of micro air-bubbles colliding with a titania plate [42, 72] are also debatable. At closer inspection, some of the experimental data for the time dependence of the separation between the apex of the bubble and the titania plate, determined from the intensity of an interference pattern, show "jumps", i.e., non-physical discontinuities (see, e.g., the curves (b) and (c) in Fig. 5 of Ref. [72]). It is possible to identify the source of these as an incorrect choice of the corresponding interference order. After correcting the experimental data for these spurious discontinuities (starting from the long time region of the curve, as this corresponds to the minimum separation and thus a clearly defined order of interference), the experimental results (b) and (c) mentioned above would lay significantly below the theoretical predictions shown there, indicative of a faster drainage and thus of a mobile (at least partially) interface, rather than of the claimed no-slip condition.

In the light of these observations, the necessity for a no slip boundary condition at a fluid-fluid interface to quantitatively explain the SFA and AFM film drainage experiments remains questionable and it seems that at this time a consistent explanation for its source is yet to be found.

4. Conclusions and outlook

We have discussed a number of theoretical and experimental challenges in the SFA and AFM studies of thin film drainage and have critically reviewed the outcomes of recent (last two decades) experimental studies. While significant progress has been made, it seems clear that the results so far cannot yet overcome all the issues outlined in Sec. 2. Therefore, we argue that conclusive SFA or AFM tests concerning the validity or otherwise of a hydrodynamic description for the drainage of liquid films with a thickness below tens of nanometers are yet to be provided. Although the theoretical modeling is greatly simplified for rigid surfaces, it is clear that the issue of a nano-rough surface reduces the likelihood of such tests being successful when solid surfaces are involved. Therefore, the most promising direction appears to be that of studying the case of deformable bounding surfaces pursued so far in, e.g., Refs. [22–25,31,33] (see also the reviews in Refs. [41, 42]. The continuing development of an interferometry enhanced AFM [73], which would allow a determination of the separation $h(r, t)$ independently of the cantilever deflection (similarly to the situation in the SFA experiments), could significantly contribute in this endeavor.

Acknowledgments

Financial support from the Australian Research Council (ARC) Linkage (grant AMSRI LP 0667828) and Discovery (grant DP 1094337) schemes is gratefully acknowledged.

References

[1] Lauga E, Brenner M P and Stone H 2007 *Microfluidics: The no-slip boundary condition*. In *Handbook of Experimental Fluid Dynamics* ed Tropea C, Yarin A and Foss J F (New York: Springer) chap 19, pp 1219 – 1240
[2] Neto C, Evans D R, Bonaccurso E, Butt H J and Craig V S J 2005 *Rep. Prog. Phys.* **68** 2859 – 2897
[3] Derjaguin B and Kusakov M 1992 *Prog. Surf. Sci.* **40** 26 – 45
[4] Fisher L R, Mitchell E E, Hewitt D, Ralston J and Wolfe J 1991 *Colloid Surf.* **52** 163 – 174
[5] Krasowska M and Malysa K 2007 *Adv. Colloid Interface Sci.* **134-135** 138 – 150
[6] Parkinson L and Ralston J 2011 *Adv. Colloid Interface Sci.* **168** 198 – 209
[7] Sheludko A 1967 *Adv. Colloid Interface Sci.* **1** 391 – 464
[8] Ivanov I B 1988 *Thin Liquid Films: Fundamentals and Applications* (New York: Marcel Dekker Inc.)
[9] Gee M L, McGuiggan P M, Israelachvili J N and Homola A M 1990 *J. Chem. Phys.* **93** 1895 – 1907
[10] Israelachvili J N 1992 *Surface Forces* (London: Academic Press)
[11] Persson B and Tosatti (Eds) E 1996 *Physics of Sliding Friction* (New York: Springer)
[12] Connor J N and Horn R G 2003 *Faraday Discuss.* **123** 193 – 206
[13] Becker T and Mugele F 2005 *Molecular Simulation* **31** 489 – 494
[14] Ducker W A, Senden T J and Pashley R M 1992 *Langmuir* **8** 1831 – 1836
[15] Honig C D F and Ducker W A 2007 *Phys. Rev. Lett.* **98** 028305
[16] Craig V S J, Neto C and Williams D R M 2001 *Phys. Rev. Lett.* **87** 054504
[17] Zhu L, Attard P and Neto C 2011 *Langmuir* **27** 6701 – 6711
[18] Guriyanova S, Semin B, Rodrigues T S, Butt H J and Bonaccurso E 2010 *Microfluid. Nanofluid.* **8** 653 – 663
[19] Vinogradova O I, Butt H J, Yakubov G E and Feuillebois F 2001 *Rev. Sci. Instrum.* **72** 2330 – 2339
[20] Vinogradova O I and Yakubov G E 2003 *Langmuir* **19** 1227 – 1234
[21] Cottin-Bizonne C, Cross B, Steinberger A and Charlaix E 2005 *Phys. Rev. Lett.* **94** 056102
[22] Dagastine R R, Manica R, Carnie S L, Chan D Y C, Stevens G W and Grieser F 2006 *Science* **313** 210 – 213
[23] Manor O, Vakarelski I U, Stevens G W, Grieser F, Dagastine R R and Chan D Y C 2008 *Langmuir* **24** 11533–11543
[24] Vakarelski I U, Manica R, Dagastine R R, Tang X, O'Shea S J, Stevens G W, Grieser F and Chan D Y C 2010 *Proc. Nat. Acad. Sci.* **107** 11177 – 11182
[25] Krasowska M, Carnie S L, Fornasiero D and Ralston J 2011 *J. Phys. Chem.* C **115** 11065 11076
[26] Israelachvili J N 1985 *J. Colloid Interface Sci.* **110** 263 – 271
[27] Nam P S 1986 *Tribophysics* (New Jersey: Prentice Hall)
[28] Zhu L, Attard P and Neto C 2011 *Langmuir* **27** 6712 – 6719
[29] Honig C D F and Ducker W A 2007 *J. Phys. Chem.* C **111** 16300–16312
[30] Vinogradova O I 1995 *Langmuir* **11** 2213 2220
[31] Miklavcic S J, Horn R G and Bachmann D J 1995 *J. Phys. Chem.* **99** 16357 – 16364
[32] Pushkarova R A and Horn R G 2008 *Langmuir* **24** 8726 – 8743
[33] Manica R, Connor J N, Dagastine R R, Carnie S L, Horn R G and Chan D Y C 2008 *Phys. Fluids* **20** 032101
[34] Hartley P G, Griesser F, Mulvaney P and Stevens G W 1999 *Langmuir* **15** 7282–7289
[35] Gillies G, Prestidge C A and Attard P 2001 *Langmuir* **17** 7955–7956
[36] Nguyen A, Nalaskowski J and Miller J 2003 *Minerals Engineering* **16** 1173–1181
[37] Chan D Y C, Dagastine R R and White L R 2001 *J. Colloid Interface Sci.* **236** 141 – 154
[38] Dagastine R R and White L R 2002 *J. Colloid Interface Sci.* **247** 310 – 320
[39] Dagastine R R, Prieve D C and White L R 2004 *J. Colloid Interface Sci.* **269** 84 – 96
[40] Carnie S L, Chan D Y C, Lewis C, Manica R and Dagastine R R 2005 *Langmuir* **21** 2912 – 2922
[41] Chan D Y C, Klaseboer E and Manica R 2010 *Soft Matter* **7** 2235 – 2264
[42] Chan D Y C, Klaseboer E and Manica R 2011 *Adv. Colloid Interface Sci.* **165** 70–90
[43] Happel J and Brenner H 1973 *Low Reynolds number hydrodynamics* (Leyden: Noordhoff International)
[44] Borkovec M and Behrens S H 2008 *J. Phys. Chem.* B **112** 10795 – 10799
[45] Tabor R F, Wu C, Lockie H, Manica R, Chan D Y C, Grieser F and Dagastine R R 2011 *Soft Matter* **7** 8977 – 8983
[46] Bevan M A and Prieve D C 1999 *Langmuir* **15** 7925 – 7936
[47] Porus M, Labbez C, Maroni P and Borkovec M 2011 *J. Chem. Phys.* **135** 064701
[48] Henry C L, Dalton C N, Scruton L and Craig V S J 2011 *J. Phys. Chem.* C **111** 1015 – 1023
[49] dos Santos A P, Diehl A and Levin Y 2010 *Langmuir* **26** 10778–10783
[50] Levin Y, dos Santos A P and Diehl A 2009 *Phys. Rev. Lett.* **103** 257802
[51] dos Santos A P and Levin Y 2011 *Phys. Rev. Lett.* **106** 167801
[52] Mahanty J and Ninham B W 1976 *Dispersion Forces* (New York: Academic Press)

[53] Petrache H I, Zemb T, Belloni L and Parsegian V A 2006 *Proc. Natl. Acad. Sci.* **103** 7982 – 7987
[54] Hendy S C, Jasperse M and Burnell J 2005 *Phys. Rev.* E **72** 016303
[55] Henry C L and Craig V S J 2009 *Phys. Chem. Chem. Phys.* **11** 9514 – 9521
[56] Bonaccurso E, Butt H J and Craig V S J 2003 *Phys. Rev. Lett.* **90** 144501
[57] Hendy S C and Lund N J 2007 *Phys. Rev.* E **76** 066313
[58] Bazant M Z and Vinogradova O I 2008 *J. Fluid Mech.* **613** 125 – 134
[59] Feuillebois F, Bazant M Z and Vinogradova O I 2009 *Phys. Rev. Lett.* **102** 026001
[60] Belyaev A V and Vinogradova O I 2010 *Soft Matter* **6** 4563–4570
[61] Belyaev A V and Vinogradova O I 2011 *Phys. Rev. Lett.* **107** 098301
[62] Vinogradova O I and Belyaev A V 2011 *J. Phys.: Condens. Matter* **23** 184104
[63] Kunert C, Harting J and Vinogradova O I 2010 *Phys. Rev. Lett.* **105** 016001
[64] Asmolov E S, Belyaev A V and Vinogradova O I 2011 *Phys. Rev.* E **84** 026330
[65] Davis A M J and Lauga E 2009 *Phys. Fluids* **21** 011701
[66] Ohnishi S, Hato M, Zamada K and Christenson H K 1999 *Langmuir* **15** 3312 – 3316
[67] Zhu Y and Granick S 2003 *Langmuir* **19** 8148 – 8151
[68] Craig V S J and Neto C 2001 *Phys. Rev. Lett.* **87** 054504
[69] Craig V S J, Neto C and Williams D R M 2003 *Eur. Phys. J.* E **12** 71 – 74
[70] Bowles A P, Honig C D F and Ducker W A 2011 *J. Phys. Chem.* C **115** 8613 – 8621
[71] Wang G M, Prabhakar R and Sevick E M 2009 *Phys. Rev. Lett.* **103** 248303
[72] Manica R, Parkinson L, Ralston J and Chan D Y C 2010 *J. Phys. Chem.* C **114** 19421946
[73] Krasowska M 2011 *private communication*

Microparticles in Stokes Flows 2011 IOP Publishing
Journal of Physics: Conference Series **392** (2012) 012010 doi:10.1088/1742-6596/392/1/012010

Boundary conditions for creeping flow along periodic or random rough surfaces : experimental and theoretical results

Nicolas Lecoq

UMR 6634 CNRS, GPM, University of Rouen, 1 Avenue de l'Université, 76801 Saint Étienne du Rouvray, France

E-mail: nicolas.lecoq@univ-rouen.fr

Abstract. Hydrodynamic interactions between particles and walls are relevant for the open problem of specifying boundary conditions for suspension flows. The Reynolds number around a small particle close to a wall is usually low and creeping flow equations apply. From the solution of these equations, the drag coefficient on a sphere becomes infinite when the gap between the sphere and a smooth wall vanishes, so that contact may not occur. Physically, the drag is finite because of various reasons, one of them being the particle and wall roughness. Then, for vanishing gap, even though some layers of fluid molecules may be left between the particle and wall roughness peaks, it may conventionally be said that contact occurs. In this paper, we are considering the example of a smooth sphere moving towards a rough wall. The roughness considered here consist of random rough planes or parallel periodic wedges, the characteristic length of which is small compared with the sphere radius. This problem is considered both experimentally and theoretically. The motion of a millimetre size bead settling towards a corrugated horizontal wall in a viscous oil is measured with laser interferometry giving an accuracy on the displacement of 0.2μm. Several random rough planes and wedge shaped walls were used, with various wavelengths and wedge angles. From the results, it is observed that the velocity of the sphere is, except for small gaps, similar to that towards a smooth plane that is shifted down from the top of corrugations.

For the periodic wedges, the creeping flow is calculated as a series in the slope of the roughness grooves. The convergence of the series for the shift distance in term of the slope is accelerated by use of Euler transformation and of the existence of a limit for large slope. The cases of a flow along and across the grooves are considered separately. The shift is larger in the former case. Slightly flattened tops of the wedges used in experiments are also considered in the calculations. The effective theoretical shift for a sphere approaching a wall is obtained from Lorentz reciprocal theorem with an expansion for small roughness compared with the gap between the sphere and the wall. The effective shift is found to be the average of the shifts for shear flows in the two perpendicular directions. A good agreement is found between theory and experiment.

The theoretical description of the flow close to the random rough wall represents a difficult, nearly insurmountable problem except in lattice Boltzmann simulations. Statistical analysis is presented in this paper to deduce the effective shift for sand-blasted rough surfaces. To overcome the difficulties of modelling, a regular perturbation expansion is developed, and from Lorentz reciprocal theorem, the first order correction to the drag force due to random roughness is evaluated.

1. Introduction

Hydrodynamic interactions between particles and walls are relevant for the open problem of specifying boundary conditions for suspension flows. The Reynolds number around a small particle close to a wall is usually low and creeping flow equations apply. For a suspension in water or air, the assumption of a low Reynolds number is true for particles typically of the order of 100 μm or less. However, the application of creeping flow equations may lead to paradoxes. Classical solutions of these equations for a spherical particle close to a smooth wall [2, 5, 25, 29, 30] show that the drag coefficient becomes infinite when the gap between the sphere and the wall vanishes.

Consider a sphere with radius a moving with velocity v_p towards a smooth plane. Let ξa be the gap between the surfaces of the sphere and the plane. Solving Stokes creeping flow equations of fluid motion, the expression of the lubrication drag force on the sphere is obtained as linear function of the fluid velocity v_p in the form

$$F_z = -6\pi a\mu v_p f^T_{zz} \tag{1}$$

Here μ denotes the fluid viscosity. The force is along z by symmetry, as indicated by the subscript z. f^T_{zz} is the friction coefficient relating the force in the z-direction to the translational velocity in the z-direction.

The exact theoretical result for f^T_{zz} for the motion of a sphere towards the plane was derived independently by Brenner [2] and Maude [25] using the method of bipolar coordinates. A three terms expansion of this results for small gaps ($\xi \ll 1$) was later obtained by Cox & Brenner [4] and independently by Cooley & O'Neill [3] :

$$f^T_{zz} = \frac{1}{\xi} - \frac{1}{5}\log\xi + 0.9713 \tag{2}$$

A consequence of this lubrication effect is that the settling sphere would theoretically never touch the wall in a finite time. As recognized by Goldman, Cox & Brenner [11, 12], the practical observation that the sphere indeed touches the wall in a finite time is the consequence of other physical phenomena, such as short range attractive forces, like van der Waals forces. An earlier contact may also be due to surface roughness of the wall or particle. For even though a lubrication force on the scale of a bump prevents surfaces from approaching each other (using (2) in which a is replaced by the radius of the bump [31]), the gap between the bump and the nearby surface would then become of the order of the fluid molecular dimension. Therefore, even in the presence of some leftover layers of fluid molecules, it may conventionally be said that contact occurs. This physical importance of roughness provides thus a strong motivation for studying the hydrodynamics of suspensions with rough surfaces. The classical boundary condition for a viscous fluid at usual scales on a wall is the no-slip boundary condition. On the other hand it is known that for a gas at molecular scales, a slip condition applies outside a Knudsen layer.

Recently the no-slip boundary condition is also questioned for liquids at small scales for some particular systems, like for water on hydrophobic surfaces [22] and other systems in chemical physics, but this matter is still a topic for discussion [14] (for extensive reviews see Vinogradova [33], Ellis & Thompson [7] and more recently Vinogradova & Belyaev [34]).

Another configuration giving a slip condition will be considered here : for a non-plane wall (e.g. a rough or porous wall) on which a no-slip condition applies, there is an apparent slip on an equivalent plane wall. A slip condition proposed by Navier [28] becomes useful. It says that the slip velocity $\mathbf{U}_w$ on the wall is proportional to the viscous shear stress $\boldsymbol{\tau}_w$. It is alternatively written as proportional to the shear rate of the velocity profile $\mathbf{U}(z)$ on the wall:

$$\mathbf{U}_w = \frac{b}{\mu}\boldsymbol{\tau}_w = b\frac{d\mathbf{U}}{dz}|_w$$

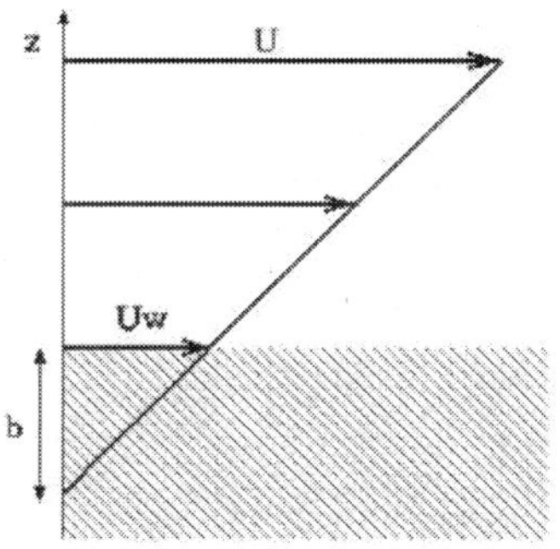

Figure 1. Sketch of the slip condition on a wall.

b appears as a length, called a "slip length" . This slip condition is equivalent to a no-slip one on a plane shifted a distance b into the wall (Fig. 1).

This article is concerned with the interactions of non-touching rough surfaces. We will study in detail the test case of a sphere approaching a rough wall. Smart & Leighton [31, 32] measured the hydrodynamic effect of the surface roughness of a sphere moving perpendicularly to a smooth wall. Some of their spheres were made rough by gluing very small spheres on their surfaces. A theoretical analysis was proposed recently by Gérard-Varret & Hillairet [10]. Here we make the reverse, that is we prepare walls with a definite roughness and the sphere roughness is small in comparison. The roughness considered in this article consists of periodic parallel wedges, the wavelength of which, λ, is small compared with the sphere radius. Experiments with wall with random roughness were realised in addition, with the objective to correlate the characteristic average value of surface roughness to the slip length. The slip length will then be derived in this typical case of a millimetre size sphere moving normal to a wall having different surface topologies and roughness on the micrometer scale.

2. Experimental setup and procedure.

The experimental setup is sketched in Fig. 2. The application of laser interferometry to the study of hydrodynamic interactions between a particle and walls in a viscous fluid has been presented extensively in preceding papers. A first version of the setup was designed in Assou *et al* [1]. A second improved version was used in Lecoq *et al* [15, 17, 18, 19], Masmoudi *et al* [23, 24] and more recently by Mongruel *et al* [26, 27]. It is then applied to the study of hydrodynamic interactions between a sphere and a rough plane. This setup is first briefly recalled in Sect. 2.1. The studied system and experimental procedure are presented in Sect. 2.2.

2.1. Description of the experimental setup.

A 15 mW He-Ne laser beam with wavelength $\Lambda = 632.8$ nm is focused at the centre of the cell with a lens L_1. It is then divided by a beam splitter into two separate beams having roughly the same power. Both beams are then deviated by successive reflections onto mirrors M, following symmetrical paths. The upper beam is reflected by the particle in motion P and the lower one is reflected by a fixed curved surface of large radius located close to the cell M_f. The two symmetrical paths have roughly equal lengths so that interferences with a good contrast are made. Both reflected beams then follow the same paths in the reverse direction, up to the separator where they combine to form interferences fringes. The resulting beam is extracted from the interferometer by the mirror M_r. It is then stopped down and focused onto the end of an optical fibre coupled to a photomultiplier.

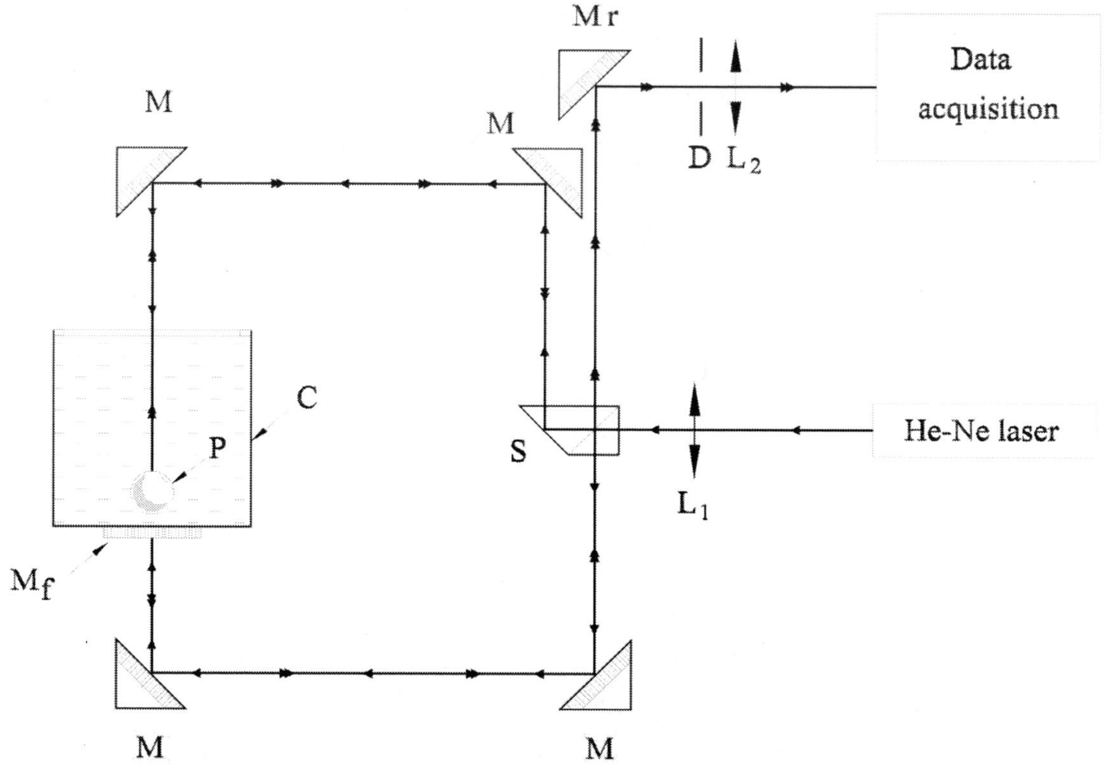

Figure 2. Experimental setup. C : cylindrical vessel; D : diaphragm; L_1, L_2 : lenses; M : mirrors; M_f : mirror reflecting the reference beam; M_r : mirror reflecting the interferences fringes; P : spherical particle; S : beam splitter.

The displacement of the particle is observed as a radial motion of the interferences fringes; it results in a shifting of the interference fringes which appear as concentric clear and dark rings. Shifting from a dark to the next clear fringe or conversely is characteristic of a displacement Δz of the particle such that

$$\Delta z = \frac{\Lambda}{4n} = 0.1126 \ \mu\text{m} \tag{3}$$

in which $n = 1.404$ is the index of refraction of the liquid.

The interferometric signal, converted to an electric one by the photomultiplier, is transformed into a digital one by a converted board installed in a PC compatible microcomputer. Thus the variation of intensity is obtained versus the time. The successive times at which the signal pass through extremi are then obtained. The difference between two such times is the elapsed time interval Δt for the particle to move a distance Δz (eq. 3). The particle velocity is then simply calculated as

$$v_{exp} = \Delta z / \Delta t$$

Thus a pratical continuous set of particle positions and velocities can be recorded. The accuracy in the displacement is roughly equal to Δz and even a little smaller. The time is recorded with the precision of the data acquisition system, of the order of 1 μs.

2.2. Studied systems.

The spherical particle of radius a is sedimenting along the centreline of the close container filled up with viscous oil. The particle used in the experiments is a steel ball manufactured by SNR with a mass density of around $\rho_p = 7800\ kg.m^{-3}$ and a diameter of 6.35 mm. Departure from sphericity is small, typically less than 0.5 μm and the arithmetic roughness R_a as indicated by the manufacture is 0.013 μm.

This vessel (with a diameter of 50 mm and a height of 40 mm) is made of altuglas and is closed at its top by a plane glass window of optical quality.

The rough surface to be studied is placed on the inside of the lower plane end wall. Several rough surfaces were used. The first series of surfaces have periodic parallel groves with height and wavelength of the order of 100 μm. A scanning electron micrograph (FEG-SEM ZEISS 1530 XB) of one surface is shown in Fig. 3. It is apparent that, due to machining, the top and bottom are not sharp wedges, but are truncated as depicted schematically in Fig. 4. The dimensions of all surfaces used in the experiment are given in table 1 [18].

Figure 3. Scanning electron micrograph of a machined wall with wedges

Table 1. Data for various profiles (λ is the wavelength and the non-dimensional top T and bottom B are defined in Fig. 4) and experimental results β_{exp} for the shift [18]. The uncertainty written for the experimental results corresponds to a range of four different measurements. The $\delta\beta$ uncertainty corresponds to the unknown final touching position of the sphere.

	α (^{o})	λ (μm)	$T\frac{\lambda}{2\pi}$ (μm)	$B\frac{\lambda}{2\pi}$ (μm)	$\beta_{exp}\frac{\lambda}{2\pi}$ (μm)	$\delta\beta\frac{\lambda}{2\pi}$ (μm)
$\mathcal{A}$	90	100	7	3	9±1	0.3
$\mathcal{B}$	90	206	28.5	17	15.5±1	1.2
$\mathcal{C}$	90	295	8	25.2	26.5±2	3.2
$\mathcal{D}$	60	100	10	11.3	10.1±1	0.3
$\mathcal{E}$	60	200	15	9.6	20±1	1.3

The other rough surface used is a random rough plane wall as it shown in Fig. 5. It consists in a brass surface that has been sand-blasted with a 20 μm grain size. A 1.0 mm × 1.0 mm area

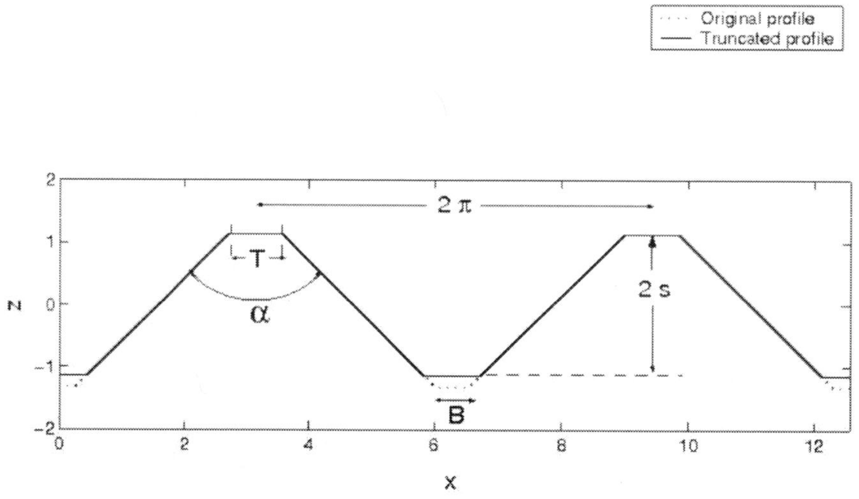

Figure 4. Profile $\mathcal{B}$. All lengths are made non-dimensional by using $\lambda/(2\pi)$ as a reference length. The bottom length B is truncated and replaced by the larger top length T. Similar approximations are made for the other profiles.

was scanned at every 5 μm in both x and y directions using a stylus-type profilometer with a 90° cone diamond tip with a radius of 5 μm. Scans were repeated 5 times with different initial locations on the surface, but close to the area where the contact will occur. The combination of surface roughness using stylus profile (Fig. 9) and optical observation using scanning electron microscopy allow more accurate evaluation of surface roughness of the wall.

The measurements obtained were R_a the arithmetic roughness, R_z the mean peak to valley height and R_{max} the maximum individual peak to valley height (table 2). Sand-blasted metal surface is well-known to present fractal dimensions as it was shown by Dubuc *et al* [6], but this point will not be taken into account in this paper.

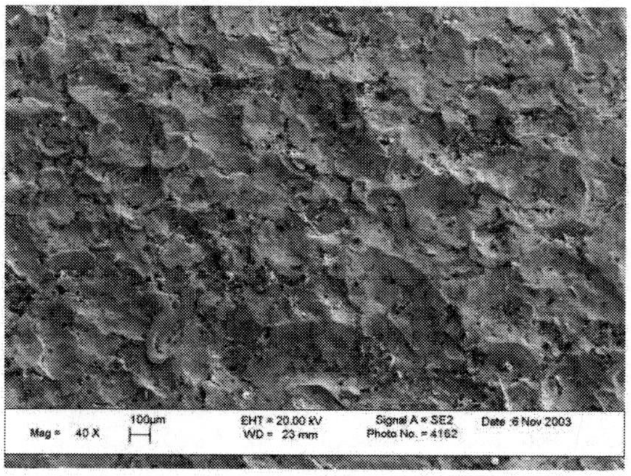

Figure 5. Scanning electron micrograph of a random rough wall

The cell contains a very viscous oil (Rhodorsyl 47V100000 manufactured by Rhône-Poulenc) with a kinematic viscosity given by the manufacturer equal to $\nu = 0.1\ m^2s^{-1}$ at $T = 25°C$.

Table 2. Data for the sand-blasted brass surface.

	R_a (μm)	R_z (μm)	R_{max} (μm)
Random Roughness	38.7	50.2	74.7

Silicon oil is chosen for its small variation of viscosity with temperature and for the physical properties. In fact, the oil is composed of dimethyl siloxane chains of various length so that even if some large chains are broken by the particle in motion, the overall distribution of chains is not much affected. The very large viscosity is chosen so that the fluid inertia effects would be quite negligible as compared to viscous effects. As a matter of fact, calculating the Reynolds number based on the sphere radius and Stokes sedimentation velocity of the particle settling in unbounded fluid at rest, we find that it is of the order of 10^{-5}. Moreover, the oil has a Newtonian behaviour up to a shear rate of the order of 100 s^{-1}. It was found by Assou *et al* [1] that this shear rate occurs in the gap when the sphere is very close to a wall, at a distance smaller than the roughness of the particles so that the non-Newtonian effects can be neglected altogether.

This oil has the required optical qualities. The refraction index of the oil at the wavelength $\lambda = 632.8\ nm$ is $n = 1.404$ and varies little with the shear rate. Moreover, the light absorption is negligible.

3. Velocity curve analysis

The contact position for which the particle velocity vanishes is clearly visible on the signal curves displayed in Fig. 6 (b). Note that the sphere may be rolling down around rugosities after this first contact and eventually come to rest in positions below the origin. A typical signal for a sphere arriving on a smooth plate is displayed for comparison in Fig. 6 (a).

The gap d between the particle and the wall is defined from this origin and reconstructed at the end of each experiment. We then define $\xi = d/a$ as the non-dimensional distance between the sphere and the wall. When the wall is smooth, the sphere velocity normalised by its Stokes velocity decays slowly with ξ in the lubrication regime and vanishes at the "contact" (Fig. 7 and Fig. 8). Note that the scales in ξ are different in Fig. 7 and Fig. 8 due to maximum roughness height which is larger for the groves than for the sand-blasted wall.

On the opposite, when the sphere approaches a rough wall, "far from the contact", the velocity decreases as previously and then there is variation in velocity and contact occurs suddenly between the sphere and the wall, that is the velocity falls to zero. In Fig. 7, we shifted the curve for the velocity obtained with the plane smooth wall by a quantity $\Delta\xi$ so that both curves matched in the far field, that is in the region where the gap ξa is an order of magnitude larger than the wavelength λ of corrugations. Practically, we used the following matching ranges for the profiles $\mathcal{A}, \mathcal{B}, \mathcal{C}, \mathcal{D}, \mathcal{E}$ described in table 1: $\xi \in [0.6, 0.8], [0.7, 0.8], [0.8, 1.0], [0.6, 0.8], [0.7, 0.8]$, respectively. For clarity, only the final part of the motion of the particle is depicted on Fig. 7. The matching ranges are limited to the far field. In addition this range of distances is yet sufficiently small so that the influence of the other walls of the container is not important. Indeed, we found that the hydrodynamic interaction with the other walls becomes important only for $\xi > 3$ [19]. If contact occurs when the sphere is touching the top of rugosities (cf Fig. 4), then by definition of β the dimensionless shift in Fig. 7 is $\Delta\xi = \beta\lambda/(2\pi a)$. Now, depending upon the initial position of the released sphere, it may also touch the rugosities after having penetrated by a small distance into the wedges. The shift $\Delta\xi$ then is smaller than $\beta\lambda/(2\pi a)$ by

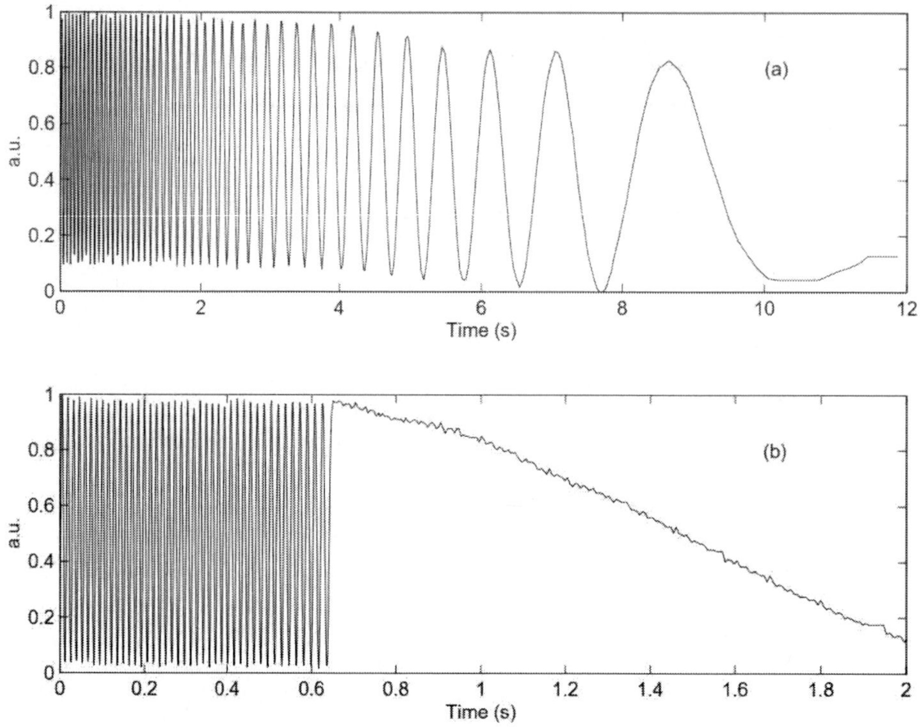

Figure 6. Example of recorded signals for the motion of the sphere arriving close to a smooth wall (a) and a rough one (b). The signal unit and origin of times are arbitrary. From a maximum to a minimum of the signal (or conversely), the sphere moves a distance $\delta z = \Lambda/4n = 0.112$ μm.

a quantity which may be estimated with some geometry to be at most

$$\delta\xi \simeq \frac{\lambda^2}{8a^2}\left(1 - \frac{T}{2\pi}\right)^2$$

leading to a variation in β of $\delta\beta = \delta\xi\, 2\pi a/\lambda$.

Results for the experimental dimensional shift $\beta_{exp}\,\lambda/(2\pi)$ (in μm) obtained for the various profiles are presented, together with the experimental errors, in table 1. We also added the dimensional variation $\delta\beta[\lambda/(2\pi)]$ for comparison. The error on the shift is larger for profile $\mathcal{C}$, since because of a larger λ, we have to match in the region $\xi \in [0.8, 1.0]$ that is farther away from the wall; the sphere falls faster and the time rate of our data acquisition system then appears a little too slow, thereby decreasing the precision.

In Fig. 8, we shifted the experimental curves (b),(c) and the theoretical ones (d), (e) for the normalised velocity by a quantity $\Delta\xi$ as previously so that curves matched in the far field, that is in the region where the gap ξa is an order of magnitude larger than the typical arithmetic roughness R_a. It is clearly observed that, depending upon the initial position of the released sphere, various behaviours are obtained. Contrary to the previous case, for the same random rough surface, the reference is chosen as the theoretical solution of Brenner [2] and Maude [25] instead of the contact between the sphere and the wall. $\Delta\xi$ is then the normalised distance between the origin of ξ and the curve (b) or (c). The curves (b) and (c) correspond to the observed extreme by repetition of the experiment (200 experiments were carried out to obtain significant statistic). The curves (d) and (e) correspond to the results of two calculations obtained by

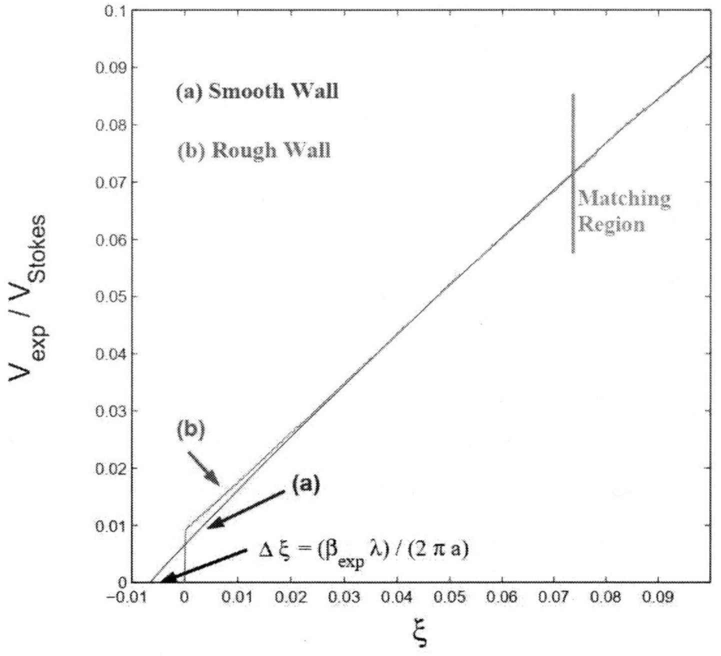

Figure 7. Example of normalized velocity v_{exp}/v_{Stokes} versus the non-dimensional gap ξ for a sphere arriving on a smooth (a) and wedges wall with profile $\mathcal{B}$ (b). A small part of the matching region is indicated.

changing the origin in the horizontal plane.

It should be noted that the curve (a) in both Figs. 7 and 8 which corresponds to an experimental measurement, is perfectly superimposed by Maude [25] and Brenner [2] theoretical solution as it was shown previously by Lecoq *et al* [15, 19].

4. Theory for the shear flow close to a periodic wall

The theory for shear flow close to a periodic wall was detailed in [18]. Only the main results are presented here.

The angles of wedges were not sharp (see Fig 4 and data in the table 1). For modelling, the original bottom part B is replaced by a length equal to the top one T . This truncated profile can then be modelled with a Fourier series. The involved error is negligible since a slack region (thus with little effect) is expected close to the bottom. Then let $2s$ be the dimensionless profile height (s may be considered as a slope) and $Z = sf(x)$ be the normalized profile, with $-1 \leq f(x) \leq 1$. For instance a Fourier series on the following form could be chosen :

$$f(x) = \sum_{m=0}^{M} f_{2m+1} \cos[(2m+1)x] \quad (4)$$

with $M = 3$ is sufficient to give a good description of profile $\mathcal{B}$ represented in Fig. 4 and values calculated later use this representation (Nevertheless, calculations were also done with $M = 4$ and 5 for comparison).

Let us define some characteristic quantities for the truncated profile. Let parameter p be the

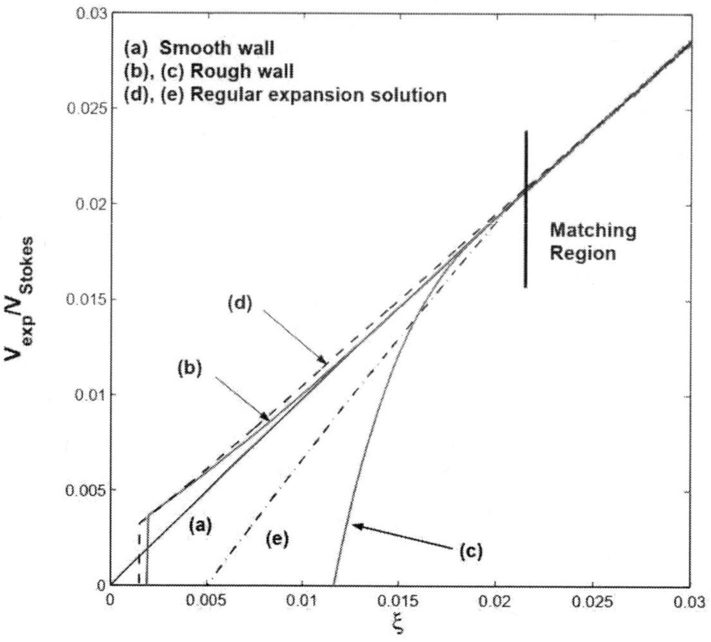

Figure 8. Example of normalized velocity v_{exp}/v_{Stokes} versus the non-dimensional gap ξ for a sphere arriving on a smooth (a) and random rough wall (b) and (c). The matching region is indicated. Curves (d) and (e) correspond to the theoretical solution obtained using a regular perturbation expansion for a sphere moving towards the rough plane wall.

ratio of lengths of the top part T to an inclined section of the truncated profile. From Fig. 4,

$$p = \frac{T}{\pi - T}$$

The dimensionless height of the truncated profile is

$$2s = (\pi - T)\cot\frac{\alpha}{2} = \frac{\pi}{1+p}\cot\frac{\alpha}{2}$$

Close to the profile, the flow due to the sphere motion is practically a shear flow. The disturbed flow due to the corrugations satisfies the Stokes equations for creeping flow. Due to the linearity of Stokes equations, the flow in any direction along the wall is solved as the sum of a flow along wedges and a flow perpendicular to them. The flow velocity along wedges satisfies simply the Laplace equation. Following [13], it is searched as a series

$$v(x,z) = Z + d_0 + \sum_{n=1}^{\infty} d_n e^{-nz}\cos nx. \tag{5}$$

The no-slip boundary condition then is applied on the profile $Z = sf(X)$.

$$0 = sf(x) + d_0 + \sum_{n=1}^{\infty} d_n e^{-nsf(x)}\cos nx. \tag{6}$$

Eq. 5 and Eq. 6 show that the equivalent no-slip plane wall is located at $Z = d_0$ so that the dimensionless slip length (distance from top of corrugations to equivalent slip plane) is $\beta = s + d_0$.

The solution for the coefficients d_n is searched by expanding them as series in s. As for the slip length, the coefficient of interest is :

$$d_0(s) = \sum_{n=0}^{\infty} a_{0,n} s^{2(n+1)}. \tag{7}$$

The convergence of series was accelerated by various techniques, so as to obtain a solution valid also for large slopes s. The results for the theoretical slip length for the flow along wedges $\beta_{th\parallel}$ are shown in the table 3.

Table 3. Comparison of β_{exp} (experiment) with β_{th} (theory) for various profiles [18].

	p	s	ζ	β_{exp}	$\beta_{th\parallel}$	$\beta_{th\perp}$	β_{th}
$\mathcal{A}$	0.16	1.35	0.65	0.57±0.04	0.72	0.40	0.56
$\mathcal{B}$	0.38	1.14	0.57	0.47±0.02	0.60	0.34	0.47
$\mathcal{C}$	0.057	1.49	0.69	0.68±0.04	0.81	0.46	0.64
$\mathcal{D}$	0.25	2.18	0.83	0.63±0.04	0.80	0.39	0.60
$\mathcal{E}$	0.18	2.31	0.84	0.63±0.04	0.86	0.41	0.63

The flow perpendicular to wedges satisfies the full Stokes equations. It was also searched as a series [13], albeit more complicated than for the Laplace equation. Coefficients in the series were also expanded as series in s. Results for the slip length $\beta_{th\perp}$ normal to wedges are also shown in the table 3. The fluid does not penetrate so much in the direction parallel to wedges as in the normal direction; the slip length then is smaller.

It was shown using Lorentz reciprocal theorem [18] that the slip length due to a shear flow at any angle to the wedges is simply obtained as the average of slip lengths in perpendicular directions:

$$\beta_{th} = \frac{1}{2}(\beta_{th\parallel} + \beta_{th\perp}) \tag{8}$$

Theoretical results for β_{th} are found to be close to the experimental ones β_{exp} within the experimental error bars. For profiles $\mathcal{A}$, $\mathcal{B}$ and $\mathcal{E}$, they are even quite close to the average experimental values. For profiles $\mathcal{C}$ and $\mathcal{D}$, β_{th} is close to the lower bound of the experimental error bar for β_{exp}. For profile $\mathcal{C}$, the process to obtain the β_{exp} involved a matching far from the wall, in the region where the precision is lower. It is also observed that for both profiles $\mathcal{C}$ and $\mathcal{D}$, the bottom flat part of the profile presented in table 1 is such that the dimensionless value B is larger than for the other profiles. Recall that the theoretical value β_{th} was calculated using the assumption that the profile is symmetrical, thus adjusting the bottom flat part B to the upper flat part T. Then, the flow in the lower part of the groove may be responsible for the slightly lower quality of the matching of the theoretical model.

5. Theory for the flow close to a random rough wall

5.1. Discussion on the normalised velocities

Fig. 8 present the two extreme behaviours of the normalised velocity (curves (b) and (c)) as the sphere approaches the random rough wall. The first one given by the curve (b) is very similar to the observed behaviour of a sphere approaching a wedge-shaped wall. Far from the wall, the velocity decreases as if the equivalent plane is smooth, and suddenly, the velocity falls to zero. On the opposite, the curve (c) shows a slow decrease of the velocity in a continuous way. This behaviour is not observed in the case of periodic wedges.

In the far field, curves (b) and (c) are perfectly superimposed to curve (a) which corresponds to the arrival of the sphere on a perfectly smooth wall. However, the shifts are different. It is seen in Fig. 9 that the random surface present bumps and valleys of various sizes. The characteristic parameters are given in table 2. The local curvature of the obstacle can strongly modify the sphere velocity at it approaches the obstacle. This is because the flow of liquid out of the region near the contact point is responsible for the drag force; and this liquid flow is strongly affected by the curvature of the obstacle.
We present here a simplified analysis on the role of the local curvature on the last part of the curves (b) and (c). The experimental shifts cannot be deduced from this analysis.
When the surface is locally convex, the outside flow is made easier and the drag decreases; the reverse happens when the surface is locally concave. It is also seen in Fig. 8 that the case of a smooth plane obstacle is intermediate between the locally convex and concave surfaces, as expected.

Consider then a sphere with radius a moving with velocity v_{exp} towards a spherical obstacle at rest (Fig.9). Let $|a_2|$ be the radius of the obstacle and let $\kappa = -a/a_2$. By definition, $a_2 > 0$, $\kappa < 0$ for a convex obstacle, and $a_2 < 0, \kappa > 0$ for a concave obstacle. For a curve obstacle, it was demonstrated by Lecoq *et al* [17] that the inverse of the normalised velocity v_{exp}/v_{Stokes} follows the form :

$$f_{zz}^T = \frac{1}{(1-\kappa)^2}\left[\frac{1}{\xi} - \frac{1}{5}\log\xi + 0.9713 + \frac{6\kappa}{5}\log\xi + O\left(K(\kappa) - 0.9713\right)\right] \tag{9}$$

The first three terms in brackets are the expansion for small gaps ($\xi \ll 1$) obtained by Cox & Brenner [4] and independently by Cooley & O'Neill [3] for the motion of a sphere towards a plane. They are equivalent to the expansion in Eq. 2.

The fourth term is small as compared to $1/\xi$, even for large value of κ.
The function of $K(\kappa)$ in the fifth term was tabulated by Jeffrey [8] and Jeffrey & Onishi [9] for the two spheres problem.

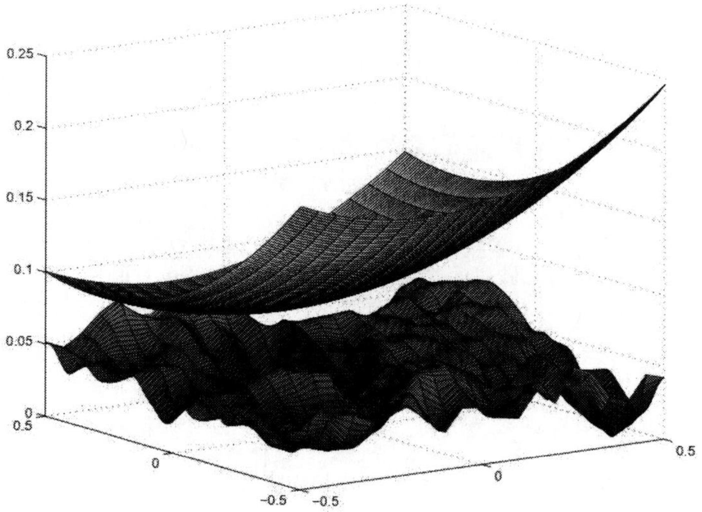

Figure 9. A large sphere approaching the random rough wall (scales are conserved). The description of the rough surface was obtained with a profilometer. Distances are indicated in millimetres.

In the present case, a rough estimation shows that the fifth term in brackets is $O(5 \times 10^{-2})$ that is very small as compared to $1/\xi$ so that this last term can be neglected.

Thus it is seen that the three-term expansion of the friction factor for a curved (convex or concave) obstacle with a small radius of curvature (smaller than the one of the moving sphere) can be obtained by multiplying the corresponding expansion for the plane by $1/(1-\kappa)^2$, that is :

$$f_{zz}^T = \frac{1}{(1-\kappa)^2}\left[\frac{1}{\xi} - \frac{1}{5}\log\xi + 0.9713\right] \tag{10}$$

From Eq. 10, the extreme behaviours presented on Fig. 8 could be explained, but the shift $\Delta\xi$ cannot be deduced precisely. When the sphere approaches the random rough wall, far from the contact, the velocity decreases as in the case of a smooth wall. The Maude [25] and Brenner [2] theory applies perfectly. When the dimensionless distance is of the order of the normalised maximum roughness height, two cases can be described :

- the sphere approaches a bump (convex peak); in that case the ratio κ takes a negative value, and the friction factor given by Eq. 10 decreases; it means that the normalised velocity of the particle increases and the final part of curve (*b*) before contact in Fig. 8 is then obtained.
- the sphere approaches a valley; the κ ratio take a positive value so that, using Eq. 10, it can be deduced an increase of the friction factor and conversely a decrease of the normalised velocity. During its downward motion, the sphere expels fluid located in the lower part of the valley; the motion of the sphere towards a valley is much slower than the one towards a plane. Some similar results were obtained by Masmoudi *et al* [23] for the lubrication motion of a sphere in a conical vessel.

5.2. Regular perturbation solution

Lecoq and Feuillebois [16] using a regular perturbation expansion obtained the contribution to the drag force due to specific roughness distributed on the plane wall (gaussian bump or valley). It is applied in this paper to the case of the random rough wall.

Consider a smooth sphere S of radius a and a rough wall W represented in cylindrical polar coordinates (r, θ, z) by :

$$z = a\epsilon f(r,\theta) \tag{11}$$

where $\epsilon \ll 1$ and $f(r,\theta)$ a normalised function describing the rough wall. The z axis is passing through the sphere center. The sphere is moving towards the wall with velocity $\mathbf{U}$ and the problem is the determination of the additional drag force due to roughness on the sphere.

The flow field is expanded for small ϵ ; the fluid velocity and stress tensor are :

$$\begin{aligned} \mathbf{v} &= \mathbf{v}_0 + \epsilon\mathbf{v}_1 \\ \boldsymbol{\sigma} &= \boldsymbol{\sigma}_0 + \epsilon\boldsymbol{\sigma}_1 \end{aligned} \tag{12}$$

The unknown drag force on the sphere is written $\boldsymbol{\mathcal{F}}$. It is searched as an expansion in ϵ :

$$\boldsymbol{\mathcal{F}} = \boldsymbol{\mathcal{F}}_0 + \epsilon\boldsymbol{\mathcal{F}}_1 \tag{13}$$

The boundary conditions are expanded for small ϵ :

- on sphere S : $\mathbf{v} = \mathbf{U}$ gives $\mathbf{v}_0 = \mathbf{U}$ and $\mathbf{v}_1 = 0$
- on wall W : $\mathbf{v}(a\epsilon f(r,\theta)) = 0$ gives

$$\mathbf{v}_0(a\epsilon f(r,\theta)) + \epsilon\mathbf{v}_1(a\epsilon f(r,\theta)) = 0$$

which expanded in Taylor series gives :

– to order ϵ^0 : $\mathbf{v}_0(0) = 0$

– to order ϵ^1 : $\mathbf{v}_1(0) = -af(r,\theta)\frac{\partial \mathbf{v}_0}{\partial z}(0)$

The solution for $\mathbf{v}_0$ is that for a smooth sphere moving perpendicular to a smooth wall. The velocity $\mathbf{v}_0$ was calculated by Brenner [2] and independently by Maude [25]; the associated stress tensor is difficult to obtain from their solution in bipolar coordinates. In the lubrication regime, both $\mathbf{v}_0$ and $\boldsymbol{\sigma}_0$ are well known. We consider here the situation in which $\mathbf{v}_0$ and $\boldsymbol{\sigma}_0$ are known, as well as the force $\boldsymbol{\mathcal{F}}_0$.

In principle it would then be necessary to solve the Stokes equations for the velocity $\mathbf{v}_1$ and associated pressure p_1 with the above boundary conditions. However, a short-cut towards the calculation of the drag may be taken by using the Lorentz reciprocal theorem [20, 21]. Apply this theorem using the two flow fields $(\mathbf{v}_0, \boldsymbol{\sigma}_0)$ and $(\mathbf{v}_1, \boldsymbol{\sigma}_1)$ between the sphere and the plane :

$$\int_{S+W} \mathbf{v}_0 \cdot \boldsymbol{\sigma}_1 \cdot \mathbf{n}\, dS = \int_{S+W} \mathbf{v}_1 \cdot \boldsymbol{\sigma}_0 \cdot \mathbf{n}\, dS \tag{14}$$

where $\mathbf{n}$ is the unit vector normal to the surface and pointing, say, into the fluid. Contribution from the surface at infinity vanishes.

We calculate separately each integral on each surface in Eq. 14 :

$$\int_S \mathbf{v}_0 \cdot \boldsymbol{\sigma}_1 \cdot \mathbf{n}\, dS = \mathbf{U} \cdot \int_S \boldsymbol{\sigma}_1 \cdot \mathbf{n}\, dS = \mathbf{U} \cdot \boldsymbol{\mathcal{F}}_1$$

where $\boldsymbol{\mathcal{F}}_1$ is precisely the term of the force that we are looking for.

$$\int_S \mathbf{v}_1 \cdot \boldsymbol{\sigma}_0 \cdot \mathbf{n}\, dS = 0$$

$$\int_W \mathbf{v}_0 \cdot \boldsymbol{\sigma}_1 \cdot \mathbf{n}\, dS = 0$$

$$\int_W \mathbf{v}_1 \cdot \boldsymbol{\sigma}_0 \cdot \mathbf{n}\, dS = -a \int_{z=0} f(r,\theta)\, \frac{\partial \mathbf{v}_0}{\partial z}(0) \cdot \boldsymbol{\sigma}_0 \cdot \mathbf{n}\, dS$$

Then from Eq. 14 the contribution to the drag force on the sphere along z is found to be

$$\mathcal{F}_1 = -\frac{a}{U} \int_{z=0} f(r,\theta)\, \frac{\partial \mathbf{v}_0}{\partial z}(0) \cdot \boldsymbol{\sigma}_0 \cdot \mathbf{n}\, dS \tag{15}$$

Note that the contribution to the drag force on the sphere $\boldsymbol{\mathcal{F}}_1$ is not necessarily perpendicular to it although $\mathbf{U}$ is. This depends on the type of wall roughness. So, it would be also possible to obtain an expression for the component of the force parallel to the wall. However, we expect this component to be small since the problem is nearly axisymmetric, and we will not consider this problem here.

A simplified version of the expression (15) of $\mathcal{F}_1$ can be obtained in the lubrication regime. Locally in the small gap between the sphere and the wall the sphere is described by the paraboloid :

$$z = a\xi + \frac{1}{2}\frac{r^2}{a} = l, \text{ say,} \tag{16}$$

where

$$\xi \ll 1$$

We assume furthermore that the roughness is small as compared with the gap :

$$\epsilon \ll \xi \ll 1$$

The order ϵ^0 problem is here the classical problem of a smooth sphere moving toward a smooth wall and at a small distance from it. The classical result for the pressure is obtained from Reynolds equation as :

$$p_0 = -\frac{3a\mu U}{l^2}$$

and from that solution the fluid velocity along r has the following parabolic profile :

$$v_{0r} = \frac{1}{2\mu}\frac{dp_0}{dr}l^2(\eta^2 - \eta)$$

where $\eta = z/l$. The velocity along z is obtained from the continuity equation :

$$v_{0r} = \int_0^z \left(-\frac{1}{r}\frac{\partial(rv_{or})}{\partial r}\right) dz$$

The classical result for the drag force on the sphere is

$$\mathcal{F}_0 = \int_S \boldsymbol{\sigma}_0 \cdot \mathbf{n} \cdot \mathbf{n}\, dS \simeq \int_S -p_0\, dS = -\frac{6\pi a\mu U}{\xi} \tag{17}$$

In the gap region, r is of order $a\xi^{1/2}$ from Eq. 16 and we let

$$\rho = \frac{r}{a\xi^{1/2}}$$

After some algebra, we calculate the term in the integral in Eq. 15 :

$$\frac{\partial \mathbf{v}_0}{\partial z}(0) \cdot \boldsymbol{\sigma}_0 \cdot \mathbf{n} = 9\frac{\mu U^2}{\xi^3 a^2}\frac{\rho^2}{\left(1+\frac{\rho^2}{2}\right)^4}$$

so that the expression for the force in Eq. 15 becomes :

$$\mathcal{F}_1 = -9\frac{\mu a U}{\xi^2}\int_{\rho=0}^{\infty}\int_{\theta=0}^{2\pi} f(a\xi^{1/2}\rho,\theta)\frac{\rho^3}{\left(1+\frac{\rho^2}{2}\right)^4}\, d\rho\, d\theta \tag{18}$$

Introducing the results (17) and (18) into (13), the z component of the force on the wall is

$$\mathcal{F} = -\frac{6\pi a\mu U}{\xi}\left[1 + \frac{3}{2}\frac{\epsilon}{\xi}\int_{\rho=0}^{\infty}\int_{\theta=0}^{2\pi} f(a\xi^{1/2}\rho,\theta)\frac{\rho^3}{\left(1+\frac{\rho^2}{2}\right)^4}\, d\rho\, d\theta\right] \tag{19}$$

Alternatively, the integral will be rather calculated in Cartesian coordinates :

$$\mathcal{F} = -\frac{6\pi a\mu U}{\xi}\left[1 + k\frac{\epsilon}{\xi}\right] \tag{20}$$

where

$$k = \frac{3}{2}\int_{\tilde{x}=-\infty}^{\infty}\int_{\tilde{y}=-\infty}^{\infty} f(a\xi^{1/2}\tilde{x}, a\xi^{1/2}\tilde{y})\frac{\tilde{x}^2+\tilde{y}^2}{\left(1+\frac{\tilde{x}^2+\tilde{y}^2}{2}\right)^4}\, d\tilde{x}\, d\tilde{y} \tag{21}$$

in which we defined

$$\tilde{x} = \frac{x}{a\xi^{1/2}}, \qquad \tilde{y} = \frac{y}{a\xi^{1/2}}$$

Eqs. 20 and 21 are the central result of the calculation. The frictional force contains two contributions of different origin. The first term independent on ϵ describes the effect of each smooth surfaces on the hydrodynamical flow in the liquid film. The second one dependent on ϵ corresponds to the perturbation of the flow due to roughness on the surfaces. In this calculation, no hypothesis were made on the shape of roughness, we just assumed that the roughness is small as compared with the gap.

The previous regular perturbation expansion is applied using the random rough wall as the plane surface. The inverse of the friction factor is plotted in Fig. 8 respectively curves (d) and (e). Curve (d) corresponds to the numerical calculation with the origin in the horizontal plane on a bump whereas for curve (e), the origin is located into a valley. The curves are shifted from a value $\Delta\xi$ close to the experimental ones, so that, in the far field the solution is perfectly superimposed to the classical Brenner [2] and Maude [25] analytical result.

The continuous integral in Eq. 21 is converted into a discrete one using the mesh of the surface obtained with the profilometer as shown in Fig. 9. To perform numerical calculation, a simple trapezoidal method is used. For the typical case used in experiments, $2a = 6.35$ mm, $\epsilon = 0.0121$, numerical results for k give $k \simeq 2.5$ for the curve (d) and $k \simeq= -0.98$ for curve (e). In both cases, the normalised velocities are in satisfactory good agreement with the experiments. The main characteristics of the velocity are well described by this approach. The disagreement between experiments and calculations could come from the limitation of the surface of integration. In fact, the calculation was made with a surface limited in size (Fig. 9) whereas the integral in Eq. 21 is infinite in each directions. The sensitivity of the results to the location of the origin is clearly demonstrated by the change of sign of the k-values. From these calculations, it is difficult to estimate the slip length for the rough wall.

It should be noticed that the regular perturbation expansion is valid far from the plane, but it is a surprise that it is also valid at close distances.

The previous calculation is applied to the grooves of wavelength λ. For a typical case used in the experiment, $2a = 6.35$ mm, $\lambda = 50\mu$m, we calculate $\epsilon = 0.0157$, and for all values of ξ such that $\epsilon < \xi < 0.5$ numerical results for k give a very small number of $O(10^{-5})$. This means that the sphere "sees" an equivalent flat plate that is located at the midplane of the zigzag roughness. The regular perturbation expansion is not fully satisfactory in this case; in fact, it was demonstrated in the previous section that the equivalent flat plane is located over the midplane.

The favourable comparison between the velocities estimated from experiments and those estimated from the regular perturbation expansion demonstrates that the calculation described here provided a method for the estimate of the effective hydrodynamic slip length, that also represents the position of the equivalent smooth plane.
This calculation provides estimates of the effective hydrodynamic slip length for motion normal to the rough surface : this may not be the relevant calculation in other situations, such as the tangential motion between two rough surfaces. In most cases, however, the normal motion of particles or surfaces is of greatest importance since it is motion in this direction which will determine whether the surfaces come into contact.

5.3. Experimental histogram

The distribution of experimental shift is presented in Fig. 10. It is obtained from 200 experiments conduct with the same random rough surface, but with various initial locations of the particle. At each experiment towards the rough surface was associated an experiment towards the perfectly smooth plane wall to obtain a precise value of Stokes velocity. The following are shown : experiments (histogram) and the classical Gauss law (full line) based on the mean value and the

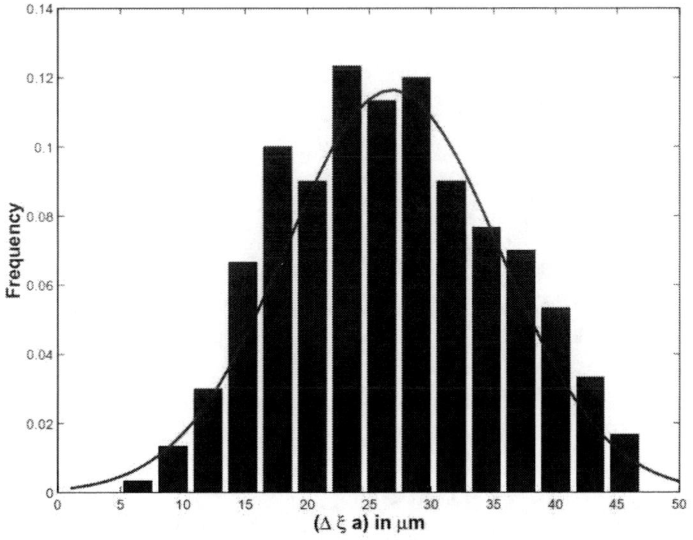

Figure 10. Histogram of the shift × the radius of the particle in μm. A Gauss law is superimposed for indication.

standard deviation from the experimental data. The mean value is estimated to 24.7 μm whereas the profilometer measurement gives 38.7 μm. Another important aspect is the low value of the shift in comparison to the one obtain with the profilometer (table 2). This may be explained by the fact that the fluid does not penetrate into the valleys; in the lower part of the valley, recirculation region may appear (dead-water area); the slip length is quite small compared with the depth and is much above the line separating the recirculation zone.

6. Conclusion

We measured the motion of a spherical particle settling towards some model surfaces made by machining various grooves or sand-blasted brass surface. Using laser interferometry, the particle displacement was obtained with an accuracy of 0.12 μm, even for very small gaps between the particle and the wall. It was then observed that, a few particle radii away from contact, the particle motion is equivalent to that towards a effective smooth plane. The position of this smooth plane (viz. the slip length) could then be obtained by matching the particle motion with that for a particle towards a plane wall.

In the case of groves, it was demonstrated that the surface is equivalent to an effective smooth plane that is displaced from the top of the corrugations by a slip length that is the average of the calculated slip lengths in the two perpendicular directions. Theoretical results for the slip length were found to be in good agreement with experiment. In the case of random roughness, various behaviour was observed, depending on the initial location of the spherical particle. The surface presents valleys or bumps, this two extreme cases result in opposite behaviours, respectively a decrease or an increase of the friction factor; by the way, the mean slip length is under-estimated for a bump and over-estimated for a valley. The favourable comparison between the velocities estimated from experiments and those estimated from the regular perturbation expansion demonstrates that the calculation described in this paper provided a method for the estimate of the effective hydrodynamic slip length, that also represents the position of the equivalent smooth plane.

More precise data could be obtained with the same interferometric technique by using other types of profiles with a well defined geometry on the microscale. A faster data acquisition system would also improve the precision in the fast part of the trajectory, when the sphere is far from the wall. This part is important for the matching allowing to determine the slip length.

It may be remarked that the present theoretical analysis for creeping flow may be used for larger Reynolds number provided that the gap between the sphere and the wall is small enough compared with the sphere radius, so that the flow field is in the lubrication regime. Moreover, from the above assumptions, the wavelength of grooves should be small compared with the gap width. Thus, this wavelength should be very small compared with the sphere radius. The measurement technique may also be used for this case, provided that the data acquisition system is fast enough to record a comprehensive set of interferometric data.

Acknowledgements

I would like to acknowledge my colleague Mr. François Feuillebois for giving me the opportunity to do my research in close collaborations. As an outstandingly open-minded interlocutor, François never hesitated to share not only his great professional competence, but even more importantly, his scientific communication skills.

The author acknowledges M.L. Ekiel-Jezewska (IPPT PAN, Warsaw, Poland) for organising the symposium "Microparticles in Stokes Flows, Symposium in honor of François Feuillebois' 65th Birthday" in Warsaw (Poland) and for providing financial support; the author acknowledges too A. Mongruel (PMMH, CNRS UMR7636, ESPCI, Paris, France) for valuable discussions. The author thanks the referees for their remarks and comments.

References

[1] Assou Y, Joyeux D, Azouni A and Feuillebois F 1991 Mesure par interférométrie laser d'une particule proche d'une paroi *J. Phys. III* **1** 315–330
[2] Brenner H 1961 The slow motion of a sphere through a viscous fluid towards a plane surface *Chem. Eng. Sci.* **16** 242
[3] Cooley M D A and O'Neill M E 1969 On the slow motion generated in a viscous fluid by the approach of a sphere to a plane or stationary spheres *Mathematika* **10** 37–49
[4] Cox R G and Brenner H 1967 The slow motion of a sphere through a viscous fluid towards a plane surface. II. Small gap widths including inertial effects *Chem. Eng. Sci.* **22** 1753–77
[5] Dean W R and O'Neill M E 1963 A slow rotation of viscous liquid caused by the rotation of a solid sphere *Mathematika* **10** 13–24
[6] Dubuc B, Zucker S W, Tricot C, Quiniou J F and Wehbi D 1989 Evaluating the fractal dimension of surfaces *Proc. R. Soc. London* **A 425** 113-127
[7] Ellis J S and Thompson M 2004 Slip and coupling phenomena at the liquid-solid interface *Phys. Chem. Chem. Phys* **6** 4928–38
[8] Jeffrey D J 1982 Low-Reynolds-number flow between converging spheres *Mathematika* **16** 106–121
[9] Jeffrey D J and Onhishi Y 1984 Calculation of the resistance and mobility functions for two unequal rigid speres in low-Reynolds-number flow *J. Fluid Mech.* **139** 261–290
[10] Gérard-Varet D and Hillairet M 2011 Computation of the drag force on a rough sphere close to a wall *Arxiv preprint arXiv:1103.0864*
[11] Goldman A J, Cox R G and Brenner H 1967 Slow viscous motion of a sphere parallel to a plane wall : I Motion through a quiescent fluid *Chem. Engng Sci.* **22** 637–651
[12] Slow viscous motion of a sphere parallel to a plane wall : II Couette flow *Chem. Engng Sci.*, **22** 653–660
[13] Hocking L 1976 A moving fluid interface on a rough surface *J. Fluid Mech.* **76** 801–817
[14] Lauga E, Brenner M P and Stone H A 2005 *Microfluidics: The no-slip boundary condition* (Handbook of Experimental Fluid Dynamics) ed C Tropea, J Foss and A Yarin (New York: Springer-Verlag)
[15] Lecoq N, Feuillebois F, Anthore N, Anthore R, Bostel F and Petipas C 1993 Precise measurement of particle-wall hydrodynamic interactions at low Reynolds number using laser interferometry *Phys. Fluids A* **5(1)** 3-12
[16] Lecoq N 1994 Étude des interactions hydrodynamiques particules-parois par interférométrie laser *PhD thesis Rouen University*

[17] Lecoq N, Feuillebois F, Anthore R, Petipas C and Bostel F 1995 Experimental investigation of the hydrodynamic interactions between a sphere and a large spherical obstacle *J. Phys. II* **5** 323-334
[18] Lecoq N, Anthore R, Cichocki B, Szymczak P and Feuillebois F 2004 Drag force on a sphere moving towards a corrugated wall *J. Fluid Mech* **513** 247–264
[19] Lecoq N, Masmoudi K, Anthore R and Feuillebois F 2007 Creeping motion of a sphere along the axis of a closed axisymmetric container *J. Fluid Mech* **585** 127–152
[20] Lorentz H A 1897 A general theorem concerning the motion of a viscous fluid and few consequences derived from it *Versl. Kon. Acad. Wet. Amst.* **5** 166
[21] Lorentz H A 1906 *Abhandl. Theoret. Phys.* **1** 23
[22] Lumma D, Best A, Gansen A, Feuillebois F, Radler J and Vinogradova O I 2003 *Phys. Rev. E* **67** 110
[23] Masmoudi K, Lecoq N, Anthore R, May S and Feuillebois F 1998 Lubricating motion of a sphere in a conical vessel *Phys. Fluids* **A 10**, 1231–33
[24] Masmoudi K, Lecoq N, Anthore R, Bostel F and Feuillebois F 2002 Accurate measurement of hydrodynamic interactions between a particle and walls *Exp. in Fluids* **32** 55–65
[25] Maude A D 1961 End effects in falling-sphere viscometer *Br. J. Appl. Phys.* **12** 293
[26] Mongruel A, Lecoq N, Wajnryb E, Cichocki B and Feuillebois F 2011 Motion of a sphero-cylindrical particle in a viscous fluid in confined geometry *Eur. J. Mech. B-Fluids* **30, 4** 405–408
[27] Mongruel A 2012 Near-wall hydrodynamic interactions between a settling sphere and a wall *J. Phys.: Conf. Ser.* **392** 012011
[28] Navier C L M H 1823 Mémoires de l'Académie Royale des Sciences de l'Institut de France **1** 414–416
[29] O'Neill M E 1964 A slow motion of viscous liquid caused by a slowly moving solid sphere *Mathematika* **11** 67–74
[30] O'Neill M E and Stewartson K 1967 On the slow motion of a parallel to a nearby wall *J. Fluid Mech* **27** 705–724
[31] Smart J and Leighton D T 1989 Measurement of the hydrodynamic surface roughness of non-colloidal spheres *Phys. Fluids* **A 1** 52–60
[32] Smart J, Beimfohrd S and Leighton D T 1993 Measurement of the translational and rotational velocities of non-colloidal sphere rolling down a smooth inclined plane at low Reynolds number *Phys. Fluids*, **A 5** 13–24
[33] Vinogradova O I 1999 Slippage of water over hydrophobic surfaces. *Int. J. Mineral Proc.* **56** 31–60
[34] Vinogradova O I and Belyaev A V 2011 Wetting, toughness and flow boundary conditions *J. Phys. Condens. Matter*, **23**

Near-wall hydrodynamic interactions between a settling sphere and a wall

Anne Mongruel

PMMH, UMR 7636 CNRS-ESPCI-University Pierre et Marie Curie (Paris 6) -University Paris Diderot (Paris 7), 10 rue Vauquelin, 75231 Paris cedex 05, France

E-mail: anne.mongruel@upmc.fr

Abstract. The near-wall dynamics of a sphere is experimentally resolved at microscale using a specially designed interferometry-related device. The millimeter size sphere, settling in a viscous oil towards a horizontal wall, is used as a reflector. The accuracy on the sphere displacement is smaller than 0.2 μm. Since its first implementation, this technique has been used to investigate various issues related to near-wall hydrodynamics. This presentation focuses on two recent developments of this device. In the first case of a sphere entering the near-wall region with a Reynolds number of order unity, it is shown how the dynamics of the sphere is modified by small inertia effects. In the second case, if a textured wall with microscopic periodic grooves is used, the velocity of the sphere increases, as compared to the velocity near a "smooth" wall. This velocity enhancement is measured accurately and correlated to the texture geometry. The results are useful for specifying effective boundary conditions at microscale, with possible applications in the field of microfluidics.

1. Introduction

The flows of suspensions are of wide interest in industrial applications and natural phenomena. The case of particles interacting with walls is very important for specifying the boundary conditions in particulate two-phase flows. Recently, these boundary-related problems have received a renewed attention with the development of microfluidic devices. However, their complete understanding is still lacking. This paper illustrates how the near-wall dynamics of a sphere can be experimentally resolved using an interferometer device (Fig. 1), in which a millimeter size sphere, settling in viscous oil towards a horizontal wall, is used as a reflector. The accuracy on the displacement of the sphere is smaller than 0.2 μm. Since its early implementation [1, 2], this technique has been used to tackle various issues related to hydrodynamic interactions with walls [3, 4] . Here, the focus is on the region near to the wall, with sphere-to-wall distance, h, restricted to a small fraction of the sphere radius, a : typically $h \leq 0.02a$. In that region, the drag on a sphere moving with velocity $V(h)$ towards the wall has been calculated exactly, in the absence of fluid inertia, as a function of dimensionless gap h/a, and reads [5] :

$$F_C = 6\pi\mu a V \left(\frac{a}{h} + \frac{1}{5}\, ln \left[\frac{a}{h}\right] + 0.9713\right) \qquad (1)$$

where μ is the fluid dynamic viscosity. Eq. (1) reflects the rapid increase in drag as the distance between sphere and wall decreases. It is only valid for $h \leq 0.3a$. For smaller gaps ($h \leq 0.01a$),

 doi:10.1088/1742-6596/392/1/012011

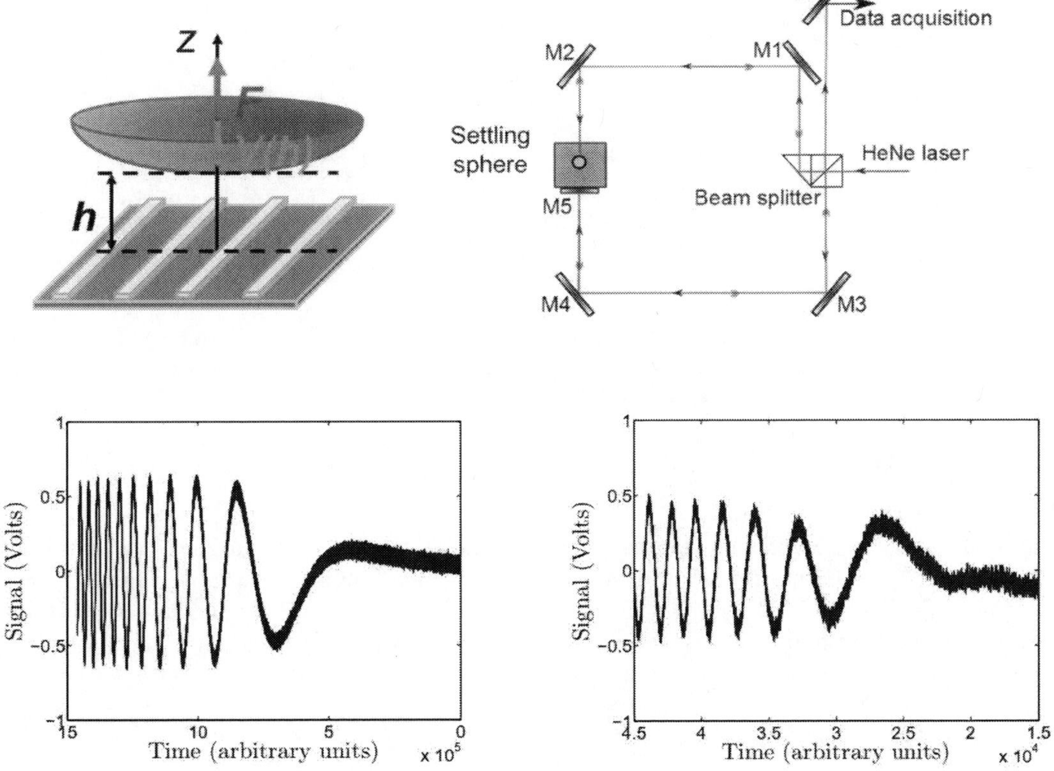

Figure 1. Sketch of the experimental set-up : sphere in the close vicinity of a surface (left), here a model striped surface, the distance h is measured from the contact with the top of the stripes. Interferometric device (right), mirror M5 is removed in case of the approach of a smooth (glass) wall. Bottom : recorded signal for a sphere approaching a smooth wall (left) and a striped wall (right) at small Reynolds number. The time unit and origin of times are arbitrary. The approach to the wall occurs from left to right of the signal-time panel.

Eq. (1) reduces to the lubrication drag force (Taylor's formula) :

$$F_T = \frac{6\pi\mu a^2 V}{h} \tag{2}$$

In settling experiments, in the absence of particle inertia, the drag force on the sphere is a constant equal to gravity forces, that is in turn also equal to the Stokes drag $F_{St} = 6\pi\mu a V_{St}$, where V_{St} is the Stokes velocity (terminal sphere velocity in an unbounded fluid). Therefore, the velocity $V_C(h)$ under the action of a constant buoyancy force given by Eq.(1) , is expected to read :

$$\frac{V_C(h)}{V_{St}} = 1/(\frac{a}{h} + \frac{1}{5}\, ln\left[\frac{a}{h}\right] + 0.9713) \tag{3}$$

with a leading term equal to h/a. Eq.(3) is experimentally verified in the reference case of a (smooth) sphere approaching a quasi smooth-wall at small Reynolds number (Fig. 2). Note that the dimensionless velocity coincides with h/a only in the very near-wall region (($h/a \leq 0.01$),

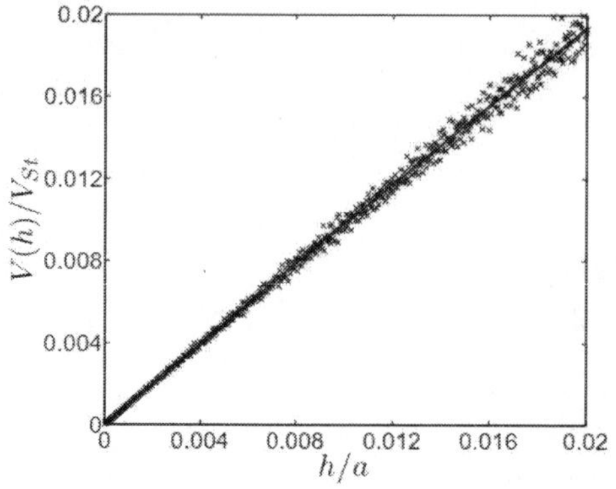

Figure 2. Dimensionless velocity versus dimensionless sphere-to-wall distance, obtained in case of a sphere approaching a smooth wall at small Reynolds number. Crosses : experimental points for a quasi-smooth wall (glass), Line : prediction from Eq.(3).

i.e the range of validity of Eq.(2).
In the following, two deviations to this reference situation will be examined. In the first case, the approach of the sphere to a (smooth) wall occurs at a Reynolds number of order unity, i.e sphere and fluid have non-negligible inertia. In the second case, a sphere (with negligible inertia) approaches a well-defined textured surface. The fluid inertia is negligible. The texture under study consists of periodic rectangular grooves, the wavelength of which is small compared to the sphere radius.

2. Materials and procedures

2.1. Fluids and particles

The sphere settles by gravity in a fluid contained in a cylindrical vessel, with a 50 or 80 mm diameter, and a 40 mm height, towards the horizontal wall of interest. The lateral walls are made with altuglass, and the top and bottom plane walls are made of glass of optical quality. The fluids are PDMS (silicone) oils of different viscosities, of the 47V Rhodorsyl series (by Rhône-Poulenc). The physical properties at 25°C, are : density $\rho = 978\,\mathrm{kg\,m^{-3}}$, and kinematic viscosity $0.1\mathrm{m^2s^{-1}}$ (resp. $10^{-3}\mathrm{m^2s^{-1}}$) for experiments at small Reynolds number (resp. at Reynolds number of order unity). Experiments are conducted at ambient room temperature. Physical properties of the oil (viscosity, refraction index) vary slowly with temperature. The oils have a Newtonian behaviour for the shear rates under study, even in the near-wall region. The particles are spherical steel balls with density $\rho_p = 7.8 \times 10^3\,\mathrm{kg\,m^{-3}}$ and diameters ranging from 5.55 mm to 14 mm. The arithmetic roughness R_a of the particles, as given by manufacturer, is 0.013 μm.

2.2. Interferometry

An interferometric technique [2, 6, 7] is used to measure the displacement of the sphere at distances from the wall smaller than 1mm. In the configuration with transparent vessel bottom, two laser beams are reflected by the top and bottom of the moving sphere. Interference fringes (circular rings) are formed, that move according to the sphere displacement. The recorded signal is the light intensity at the centre of the interference pattern. In this configuration, a signal variation from a maximum (bright fringe) to another maximum (bright fringe) is due to a displacement of the sphere equal to $\lambda/4n$, where λ is the wavelength of the laser, and n the refraction index of the suspending fluid. Here, with a silicon oil ($n = 1.404$) and a He-Ne laser beam ($\lambda = 632.8nm$), we have $\lambda/4n = 0.112\mu$m. The velocity $V(h)$ of the sphere is obtained by multiplying $\lambda/4n$ by the frequency of the signal. For the experiments with textured surfaces, however, the experimental set-up is the same as described in [6] : only one of the laser beam is reflected by the top of the moving sphere, whereas the second beam is reflected by a (fixed) mirror at the outside of the bottom container wall. Here, the interfringe becomes equal to $\lambda/2n$. In order to follow the faster sphere motion occuring at $Re \cong 1$, the data acquisition system and treatment procedure were updated (for more details, see [7]). Currently, the maximum velocity that can be detected is $100mm/s$. The signal, after opto-electronic conversion and amplification, is recorded with a high frequency electronic oscilloscope (DPO4032 from Tektronics), transferred to a PC and processed by a Matlab code for detecting its extreme values as a function of time. The spatial resolution is therefore equal to 0.112 μm (resp. 0.224 μm), for the interferences formed with two (resp. one) beam on the sphere. Fig. 1 depicts the signal evolution for a sphere arriving at a quasi-smooth wall (optical glass). The deceleration of the sphere corresponds to an increase in the period of the signal, until contact occurs. The contact position with the bottom wall ($h = 0$) is defined at the time when the period of the signal becomes very large indicating a vanishing velocity. Note that the signal-to-noise ratio deteriorates at vanishing frequency, because the low frequency limit of the oscilloscope is reached. Finally, the velocity $V(h)$ of the sphere is obtained for each value of the sphere-to-wall distance, h, that is reconstructed from the contact position (top of the roughness or textures, respectively). The measured frequency is averaged over 7 to 8 periods,except just before the contact, where no averaging is applied in order to capture the rapid velocity variations occuring in that region.

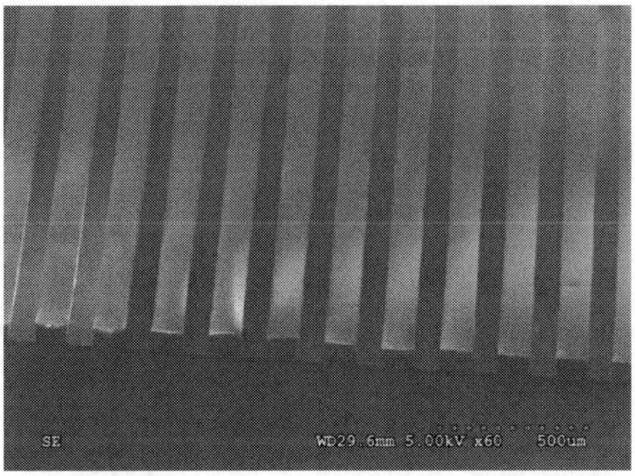

Figure 3. Scanning electron micrograph of a micro-patterned NOA surface obtained by soft lithography (dimension of the grooves : $e = 45$ μm, $L = 100$ μm, $\phi = 0.5$)

2.3. Soft lithography

Various surfaces with controlled texture geometry were fabricated in a clean room. In a first step, geometric shapes are transferred from a mask to a silicone (Si/SiO2) wafer coated with a photo-reticulable (SU8, MicroChem) resin, by exposure to UV, developing and hard-baking. The thickness of the film is controlled by spin coating. The size of the structures obtained in this first step are checked by mechanical profilometry. In a second step, a replica molding of the SU8 structures is obtained by soft lithography, using a thermo-reticulable PDMS (Sylgard 184). A last step is a replica of the PDMS molding by soft imprint of a photo-curable polymer, a thiolene based resin (NOA 81, Norland optical adhesives) on glass microscope slides (to be fixed at the bottom of the vessel). This resin has a good resistance to compression, a good adhesion to glass [8]. The wetting of NOA surfaces by silicon oil is proofed by deposition of a millimetric drop of oil : after complete imbibition of the textures by the oil, the contact angle reaches an equilibrium value lower than 30°. Thus, the surface can be considered as lyophilic : it is expected that the oil completely fills the grooves of the textures, without air-bubbles being trapped at the solid-oil interface.

3. Sphere-wall hydrodynamic interactions with small particle inertia

3.1. Position of the problem

Let us consider a sphere of mass m_p, with a density ρ_p of same order as the fluid density ρ, settling in this fluid with a velocity such that the sphere enters the near-wall region with non-negligible inertia. Particle inertia is characterized by a Stokes number, ratio of particle inertial forces to viscous forces. Here, an impact Stokes number is defined as : $St_i = m_p V_i/(6\pi\mu a^2)$. It is based on an impact velocity, V_i , the terminal velocity of a sphere in an unbounded fluid at $Re \simeq 1$. This terminal velocity can be estimated by using Oseen's correction to Stokes drag $(1 + 3Re/16)$. The corresponding impact Reynolds number, $Re_i = (2a)\rho V_i/\mu$, is then related to the impact Stokes number by : $St_i = ((\rho_p/\rho)/9)/Re_i$. This yields, for the present experimental conditions, $St_i \simeq 0.9Re_i$, so that St_i is very close to Re_i. The range of impact Stokes and Reynolds number used in this study are $0.5 \leq St_i \leq 4.6$ and $0.6 \leq Re_i \leq 5.2$, respectively. Note that, with these impact conditions, collision of the particle with the wall occurs without bouncing, as the critical value of Stokes number for bouncing is situated around $St_i \simeq 10$ ([9]).

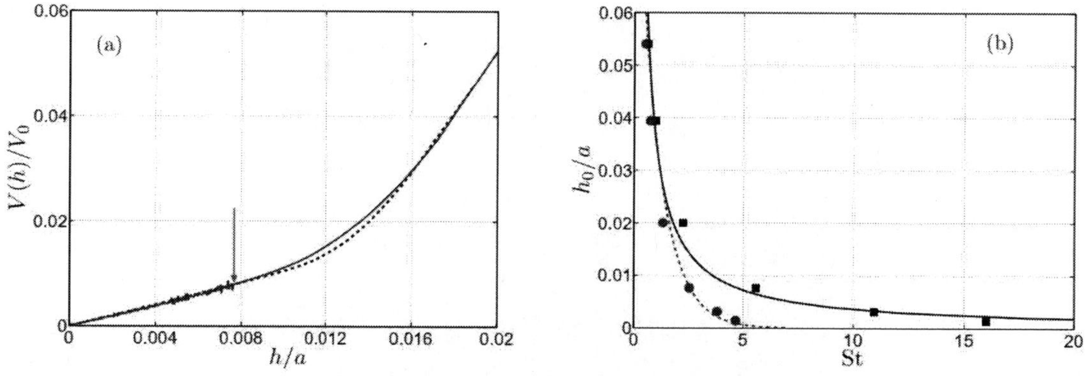

Figure 4. (a) Dimensionless velocity versus dimensionless sphere-to-wall distance, obtained in case of a sphere approaching a smooth wall at Reynolds number of order unity. Solid line : experiments at Re_i=2.8 (with a sphere of 10.50mm diameter settling towards a glass wall in a silicone oil of kinematic viscosity $10^{-3}m^2s^{-1}$). The arrow indicates the recovering of linear behaviour at the critical distance h_0/a. Dashed line : Numerical solution of Eq.(4) .
(b) Critical dimensionless sphere-to-wall distance versus Stokes number. Symbols : measurements (Stokes number based on V_0 (squares) and on V_i (circles)). Solid line : Equation $Y = 0.036/X$. Dashed line : Equation $X = ln(0.1/Y)$.

3.2. Experimental results

A typical velocity-distance curve obtained at $Re_i = 2.8$ is shown on Fig. 4(a). Two distinct regimes for the dynamics of the sphere are obtained in the near-wall region (note that the spatial scale is the same as in Fig. 2). At the entrance of the near-wall region, the dynamics is governed by inertia, and the velocity-distance curve is non-linear, reflecting the rapid deceleration of the sphere under the action of viscous forces. But at a smaller distance from the wall, the velocity has decreased sufficiently so that inertia becomes negligible and a linear regime is recovered just prior to contact, with a velocity proportional to h/a. A linear regression yields a characteristic velocity, V_0. The distance to the wall where the linear regime is recovered is denoted h_0. In practise, the value of h_0 is obtained at the distance where the relative difference between $V(h)/V_0$ and h/a becomes larger than 0.03. In the case of Fig. 4(a), the following values are obtained : V_0=51cm/s and h_0=40μm. Increasing the impact Reynolds number, and hence the characteristic velocity V_0, the spatial extension of the linear regime reduces. The smallest value of h_0 that could be measured with the experimental device is $h_0 = 10\mu m$, and was obtained for St_i=4.6, the largest impact Stokes number used in this study .

3.3. Discussion

A simple mechanical model has been developed in [7] to describe the motion of the sphere in the near-wall region, taking into account its inertia. Added mass is neglected, as it is not relevant for Reynolds numbers of order unity, except for times smaller than the time of diffusion of vorticity, which is not the case here. The difficult part in that model is that currently, no theoretical expression exists for the drag on the sphere that would account for small inertia effects of the fluid. So, an ad-hoc drag force is used, that fits to the linear regime found experimentally, in which the velocity is proportional to a characteristic velocity V_0. This drag is obtained from Eq.(2) corrected by the ratio V_{St}/V_0. Therefore, with $V(h) > 0$ pointing towards the wall, the dimensionless equation of motion reads :

$$-St_0\frac{d^2\varepsilon}{d\tau^2} = \frac{1}{\varepsilon}\frac{d\varepsilon}{d\tau} + 1 \tag{4}$$

where $\varepsilon = h/a$, τ is time divided by a/V_0 (so that $V(h)/V_0 = -d\varepsilon/d\tau$), and St_0 is a particle Stokes number based on the velocity V_0 :

$$St_0 = \frac{\rho_p V_0^2}{(\rho_p - \rho)ga} \tag{5}$$

Neglecting gravity with respect to inertia, Eq.(4) yields an exponential decreasing of velocity with distance, the same as described in earlier works dealing with elasto-hydrodynamic collision ([10, 11]):

$$\frac{V(\varepsilon)}{V_i} = \frac{V_1}{V_i} + \frac{1}{St_i}\, ln\frac{\varepsilon}{\varepsilon_1} \tag{6}$$

where V_1 is a value of $V(h)$ at the normalized distance ε_1. Note that in Eq.(6), impact parameters V_i and St_i are used.

Remarkably, a linear variation of the velocity with distance can be obtained from Eq.(4), but at a critical distance to the wall, ε_0, that is sufficiently small, so that inertia becomes negligible compared to gravity. Therefore, ε_0 verifies the condition : $\varepsilon_0 \ll 1/St_0$. Solving Eq.(4) numerically, with experimental initial conditions at the entrance of the near-wall region, gives an excellent agreement with the experimental curve, as illustrated in Fig. 4(a), not only for the linear part of the curve, as expected from the ad-hoc drag force, but also for the non-linear part, with a correct description of the sharp transition between the two regions. This indicates that the postulate made for the drag modified for fluid inertia, is correct in the entire near-wall region. On Fig. 4(b), the values of the Stokes number, St_0, are reported as a function of the spatial extension of linear region, h_0/a , obtained from experimental velocity curves. The results can be nicely fitted by the equation $St_0 = 0.036/\varepsilon_0$, indicating that the condition for ε_0 predicted by the model is correct. On the same graph, the variations of the impact Stokes number, St_i, are plotted also as a function of h_0/a, and fitted by the equation $St_i = ln(0.1/\varepsilon_0)$. The use of this logarithmic fit comes from the force balance (inertia vs lubrication drag) that holds for $1/St_0 \ll \varepsilon \ll 1$, yielding Eq.(6). The critical distance can indeed be roughly estimated as the distance ε for which $V(\varepsilon)$ vanishes in Eq.(6). The equation $St_i = ln(0.1/\varepsilon_0)$, that gives the best agreement with the data, correponds to a velocity V_1 equal to V_i at $\varepsilon_1 = 0.1$ (this impact velocity V_i resulting from the balance between drag force and buoyancy forces for $\varepsilon \gg 1$). This fit is essentially a guide for the eyes, showing that extrapolation to smaller critical distances than measured in this work, would yield an impact Stokes number that is close to the value 10, the critical value for bouncing transition [9].

4. Hydrodynamic interactions between a sphere and a textured wall

4.1. Position of the problem

Eq. (2) predicts an infinite drag force in the limit $h \to 0$. In practice, this model breaks down for various reasons [12, 13]. In particular, real surfaces are never perfectly smooth, so that the presence of roughness at the microscopic scale, modifies the drainage of the fluid squeezed between the sphere and the plane . Thus, rough surfaces make contact in a finite time. This physical importance of roughness has motivated several theoretical and experimental studies [14]. Measurement of the far-field motion of a sphere towards a periodic corrugated wall showed an increase of velocity as compared to the case of a smooth wall [6]. Recently, a decrease in hydrodynamic resistance compared to Taylor's formula was observed in the case of an aqueous film squeezed between randomly rough surfaces using an AFM-related device [15]. In the present work, experiments are performed in the case of a smooth sphere settling towards a anisotropic textured surfaces, with the textures having the form of periodic longitudinal grooves (stripes), such as depicted in Fig. 3. Geometrical parameters of the grooves (spacing δ, and wavelength L), are systematically varied, in order to cover a broad range of $\phi = \delta/L$, the area fraction of the liquid-liquid interface at the top of the grooves. The groove height e is of the order of 50 μm.

4.2. Experimental results

In Fig. 5(a), dimensionless velocities measured for spheres having different diameters , in the vicinity of the same textured surface (with following groove dimensions : $e = 42\ \mu m$, $L = 200\ \mu m$, and $\phi = 0.5$), are plotted as a function of dimensionless sphere-to-wall distance h/a. The experimental velocity is made dimensionless using the Stokes velocity V_{St}, obtained for each experiments by fitting experimental curves in the far-field to Eq.(3). Note that this procedure yields a value of V_{St} which does not need an independent measure of the fluid viscosity. The Stokes velocity obtained for the diameters 5, 7 and 8 mm is : 0.83, 1.77 and 2.32 mm/s, respectively. The case of a smooth wall is also plotted as a reference curve. The dimensionless velocity measured near the textured wall deviates from the reference curve, and is significantly above it. This increase of sphere velocity near the textured wall can be explained by the fact that the fluid, squeezed between the sphere and the wall, is able to escape through the grooves of the textures. The velocity curves in Fig. 5a, show also that the sphere decelerates abruptly on a short distance (a few number of $\lambda/2n$) before stopping by making a contact with the top of the textures (see also Fig. 1) .

In the coordinate systems of Fig.5(a), different curves are obtained for each sphere radius, with a larger mobility for the smallest radius. This indicates that here, the Stokes velocity and the sphere radius are not the relevant scales in the near-wall region, because the textures

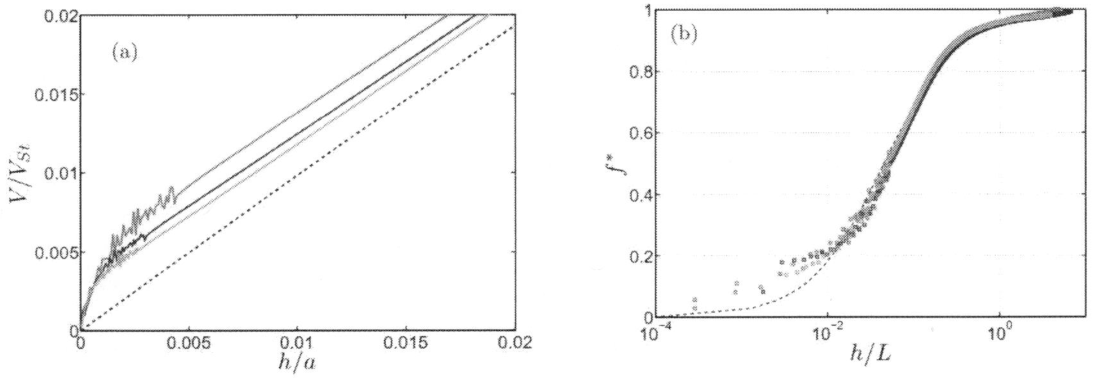

Figure 5. (a) Dimensionless velocity vs h/a , measured (continuous lines) for spheres of different diameters approaching the same striped wall ($e = 45\ \mu m$, $L = 200\ \mu m$, $\phi = 0.5$) ; from top to bottom : $2a = 5$, 7 and 8 mm . Dashed line : prediction for a smooth wall (Eq. 3).
(b) Correction to drag obtained from the same measurements (squares) as in (a) , and predicted (dashed line) by the shift model (Eq. 9) with $s = 9\mu m$.

introduce other characteristic lengths. In view of this, we evaluate the ratio :

$$f^*(h) = \frac{V_C(a,h)}{V(a,h)} \tag{7}$$

where, for a given sphere radius, $V_C(a,h)$ is the velocity at the distance h from a smooth wall (from Eq.3), and $V(a,h)$ the velocity measured at the same distance but from a textured wall. This ratio is plotted on Fig. 5(b), as a function of the distance normalized by L, the wavelength of the textures (here, $L = 200\mu m$ in all 3 experiments), in semi-Log scale. Then, all measurements obtained at different sphere diameters collapse into a single curve. As a consequence, the function $f^*(h)$ is, for a given texture geometry, only sensitive to the sphere-to-wall distance. Note that $f^*(h)$ coincides (in the absence of inertia) with the correction to drag force near a smooth wall, $F_C(a,h)$, that is introduced by the textures. In other words, $f^*(h) = F(a,h)/F_C(a,h)$, and would be obtained by measuring the drag $F(a,h)$ on a sphere moving at constant driven velocity towards the textured wall [15]. In the following, $f^*(h)$ will be interpreted in terms of correction to drag force. It is seen on Fig. 5(b) that $f^*(h)$ is close to unity at large distances, i.e the drag force is the same as for a smooth wall, however it decreases significantly as h becomes of the order of L and smaller. Finally, it vanishes at contact.
In Fig. 6 , various groove sizes are used, with approximately the same height $e = 42 - 45\ \mu m$, but different L and different ϕ. It can be seen that for a given L, the liquid area fraction ϕ is a key parameter controlling not only the shapes of the $f^*(h)$ curves, but also the intensity of drag reduction : at a given distance, drag reduction (smaller value of f^*) is promoted by a larger liquid area fraction. In Fig. 6(c) shows an example of two textures having the same ϕ but different L. The re-scaling of the distance h by the texture wavelength L gives here a remarkable collapsing of the two f^* curves.

4.3. Discussion

It is well known (for a review see [6] and references therein) that a shear flow along a periodically corrugated wall is equivalent to a shear-flow near an equivalent smooth plane, that is shifted down from the top of the corrugations. These equivalent smooth plane and associated "shift

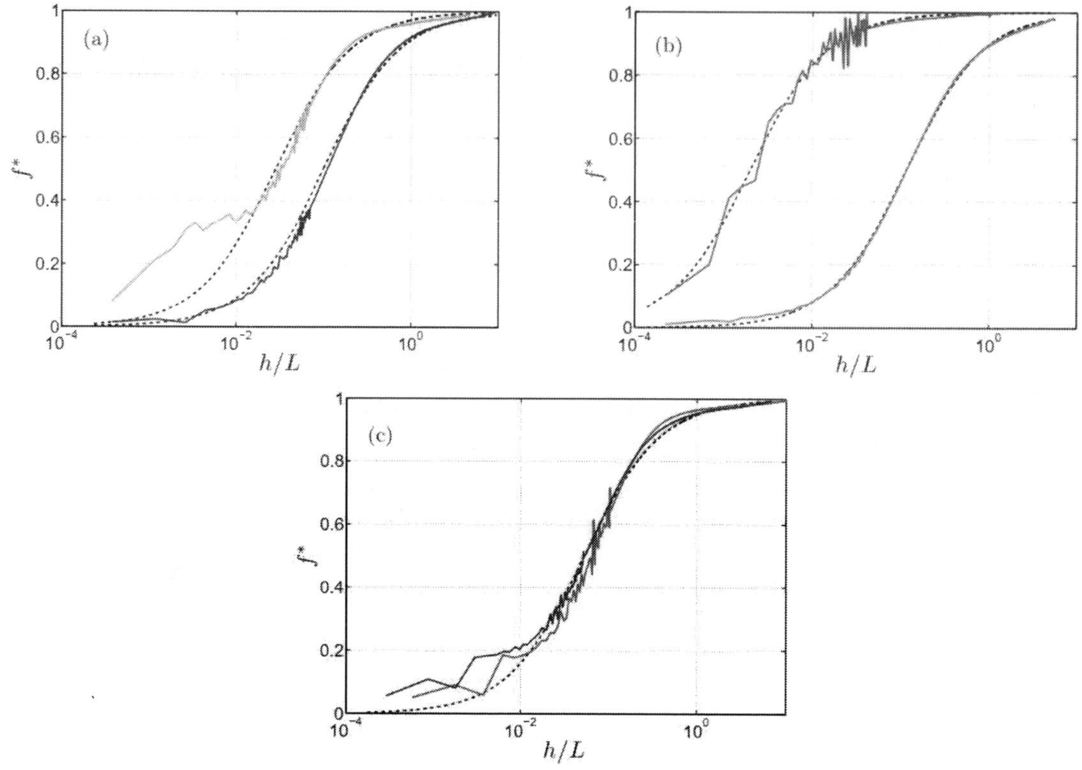

Figure 6. Correction to drag vs h/L, measured for different striped walls of nearly same height ($e = 42 - 45\ \mu m$). Continuous lines : measurements. Dashed lines : shift model (Eq.9) with values of s indicated in each case.
(a) Same $L = 150\ \mu m$, $\phi = 0.33$ (left, $s = 4.2\ \mu m$) and $\phi = 0.66$ (right, $s = 15\ \mu m$)
(b) Same $L = 250\ \mu m$, $\phi = 0.1$ (left, $s = 0.5\ \mu m$) and $\phi = 0.9$ (right, $s = 28.5\ \mu m$)
(c) Same $\phi = 0.5$, $L = 200$ and $100\ \mu m$ ($s = 9\ \mu m$ and $s = 5\ \mu m$, respectively).

length" result from averaging the flow on a scale larger than the corrugations. They are effective properties of the flow, and in that sense are distinct from the concept of hydrophobic slippage, a local property giving rise to the so-called "slip length". In the case of the film drainage between a sphere and a plane with periodic corrugations, the flow is the sum of two shear flows in the directions parallel and perpendicular to the corrugations. As a consequence, the following shift in drag force has been proved theoretically [6] :

$$F(h) = F_C(h + s) \tag{8}$$

where s is the shift length, i.e. the distance from the top of the textures at which this equivalent smooth plane is shifted from the real surface. Moreover, s has been shown to be the average of effective shift lengths in the directions parallel an perpendicular to the corrugations, respectively. These ideas were confirmed recently, and generalized to any type of texture geometry [16], using the formalism of tensorial slip [17], with a shift length obtained as the average of the eigenvalues of the slip tensor [16]. Using Eq. 8 together with Eq. 1 yields , at the leading order, the following correction to drag :

$$f^*(h) = \frac{F_C(h + s)}{F_C(h)} \cong \frac{h}{h + s} \tag{9}$$

The ratio of measured velocity to Stokes velocity, then reads :

$$\frac{V(h)}{V_{St}} \cong \frac{h}{a}\frac{h+s}{h} \cong \frac{h+s}{a} \tag{10}$$

Using spheres with different radii, Eq.(10) predicts that plotting the velocity ratio versus h/a yields parallel straight lines, with s/a corresponding to the horizontal shift along the h/a axis, the smallest radius giving the largest shift. As the curves are of slope 1, they also have an intercept s/a with the $h=0$ axis, in agreement with the experimental results of Fig. 5(a).
Now, the value of the shift length, s, depends on each texture geometry, and is obtained by fitting f^* to Eq.(9), searching the best agreement with the experimental curve in the far-field ($h >> L$). These fits are depicted as dashed lines in Fig. 5 and Fig. 6, showing good agreement with experiments at all distances, except very small ($h \leq 0.01L$). The values of s obtained by this procedure are significant compared to the groove height, remaining however smaller than it, as expected. As an example, the maximum value obtained in this study is $s = 28.5\mu m$, for the following groove dimensions $e = 42\ \mu m$, $L = 250\ \mu m$, and $\phi = 0.9$. At a fixed L, s is increasing wih increasing ϕ. The collapsing of the two f^* curves obtained for a same ϕ but two different L, indicates that s is essentially proportional to L for this value of ϕ .
The present results are currently complemented by varying systematically the size of the grooves, in order to correlate the value of the shift length s, to the parameters (surface coverage ϕ, wavelength L, depth e) of the texture. Preliminary results show that s is proportional to e at fixed ϕ and L. A comparison with theoretical predictions for the shift length is under progress.

5. Conclusion and perspectives

This paper shows the capabilities of the interferometric device to bring new insights in the near-wall dynamics of a sphere, in two situations of current interest. The results obtained can be used for a better description of effective hydrodynamic boundary conditions, that are of special relevance for systems with size reduced to micro-scales. They also open the way to further investigations. First, concerning the textured surfaces with microgrooves, a theoretical examination of hydrodynamic interactions at distances much smaller than the groove wavelength, is required. Then, using various texture geometries (e.g. pillars), would enlarge the range of experimental results for effective boundary conditions. On another hand, studying the combined effect of sphere inertia, and wall (or sphere) textures, and increasing the particle Stokes number, would possibly bring new informations on the transition towards bouncing. Finally, using non-Newtonian fluids, the modifications brought by their non-linear flow properties to the dynamic of the sphere, could be detected, and compared to recent analytical and numerical studies on the flow of non-Newtonian fluids squeezed between a sphere and a wall [18].

Acknowledgments

I thank my collaborators François Feuillebois, Samir Yahiaoui, Cyril Lamriben, and Thibault Chastel, for their highly valuable participation at various stages of this work. I am grateful to Nicolas Lecoq for technical support and helpful remarks. The support of the "PICS 4696" cooperation agreement between CNRS (France) and PAN (Poland) is gratefully aknowledged.

References

[1] Assou Y, Joyeux D, Azouni A and Feuillebois F 1991 *J. Phys. III 1* **1** 315-330
[2] Lecoq N, Feuillebois F, Anthore N, Anthore R, Bostel F and Petipas C 1993 *Physics of fluids A* **5** 3-12
[3] Masmoudi K, Lecoq N, Anthore R, Bostel F and Feuillebois F 2002 *Exp. Fluids* **32** 55-65
[4] Mongruel A, Lecoq N, Wajnryb E, Cichocki B and Feuillebois F 2011 *Eur. J. of Mech. B : Fluids* **22** 405-408
[5] Cox R G and Brenner H 1967 *Chem. Eng. Sci* **22** 1753-1777
[6] Lecoq N, Anthore R, Cichocki B, Szymczak P and Feuillebois F 2004 *J. Fluid Mech.* **513** 247-264

[7] Mongruel A, Lamriben C, Yahiaoui S and Feuillebois F 2010 *J. Fluid Mech.* **661** 229-238
[8] Bartolo D, Degre G, Nghe P and Studer V 2008 *Lab Chip* **8** 274-279
[9] Gondret P, Lance M, and Petit L 2002 *Phys. Fluids.* **14 (2)** 643-652
[10] Davis R H, Seyrassol J M and Hinch E J 1986 *J. Fluid Mech.* **163** 479-497
[11] Barnocky G and Davis R H 1988 *Phys. Fluids.* **31 (6)** 1324-1329
[12] Smart J R and Leighton D T 1989 *Phys. Fluid A* **1** 52-60
[13] Sundararajakumar R R and Koch D L 1996 *J. Fluid Mech.* **313** 283-308
[14] Lecoq N 2012 *J. Phys. : Conf. Ser.* **392** 012010
[15] Vinogradova O I and Yakubov G E 2006 *Phys. Rev. E* **73** 045302
[16] Asmolov E S, Belyaev A V and Vinogradova O I 2011 *Phys. Rev. E* **84** 026330
[17] Bazant M Z and Vinogradova O I 2008 *J. Fluid Mech.* **613** 125-134
[18] Despeyroux A and Ambari A 2012 *J. Non-Newtonian Fl. Mech.* **167-168** 38-45

Influence of wall slip in dilute suspensions

François Feuillebois[1], Néjiba Ghalya[2], Antoine Sellier[3] and Lassaad Elasmi[2]

[1] LIMSI-CNRS, UPR 3251, B.P. 133, 91403 Orsay cédex, France
[2] Laboratoire d'Ingénierie Mathématique, École Polytechnique de Tunisie, BP 743, 2078 La Marsa, Tunisia
[3] LadHyX. École Polytechnique, 91128. Palaiseau cédex, France

E-mail: francois.feuillebois@limsi.fr

Abstract.
Navier's (1823) slip condition, where the tangential velocity is proportional to the shear stress, may apply at micro-scales for a viscous liquid on walls with modified surfaces, e.g. hydrophobic ones in water. Experiments by Lumma et al (2003) show the difficulties involved in measuring such a slip.

At small scales, particle-wall hydrodynamic interactions are important. They are modeled here for a dilute suspension of spherical solid particles near a slip wall. Consider a translating and rotating sphere (on which the no-slip condition applies) in an ambient parabolic flow. Analytical solutions of Stokes equations for the various elementary flow fields were obtained (Feuillebois et al 2009, 2011) as series in bispherical coordinates. The coefficients in the series are solutions of an infinite linear system, which is solved by an extension of Thomas' algorithm, allowing to calculate a large number of terms. Accurate results are available for the force, torque and stresslet on a sphere, velocity of a freely moving sphere and diffusion tensor.

The Aris-Taylor dispersion of Brownian particles in a shear flow near a slip wall gives a large bias in the measurement of slip (Vinogradova et al, 2009). It is calculated here from the advection-diffusion equation, using the expressions for the particle velocity and diffusion tensor near a slip wall.

1. Introduction

In this paper, prepared at the occasion of the symposium for my 65th birthday, I shall present a summary of results obtained with various collaborators over the recent years, namely Olga I. Vinogradova and Dirk Lumma for the experimental part [1, 2], Hend Loussaief, Laurentiu Pasol for the theoretical part on creeping flow [3]. The paper also contains recent theoretical results [4] obtained with my present coauthors, Néjiba Ghalya, Antoine Sellier and Lassaad Elasmi. I am grateful to Maria L. Ekiel-Jeżewska and to the local committee for the splendid organisation of this symposium.

The idea of a slip boundary condition for a viscous fluid on a solid wall was originally postulated by Navier [5]. This condition is written in term of a slip velocity v_w along the wall that is proportional to the tangential stress τ_w. Alternatively, for a flow along a plane wall, v_w is proportional to the shear rate on the wall $\mathrm{d}v/\mathrm{d}z|_w$, where v is the velocity along the wall at distance z:

$$v_w = \frac{b}{\mu}\tau_w = b\left.\frac{dv}{dz}\right|_w$$

Here, μ is the fluid viscosity. The constant b is the slip length. It may be interpreted as the distance into the wall for which the extrapolated shear flow velocity (based on the shear rate at the wall) would vanish (Fig. 1. Note that in general we do not extrapolate the velocity profile itself to zero; this only happens for a linear shear flow). Thus, alternative ways of representation of the slip condition for a shear flow are the slip velocity v_w and the slip length b.

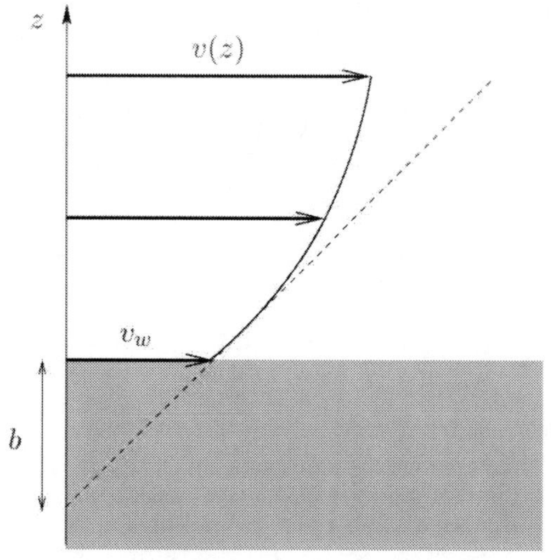

Figure 1. Definitions of the slip velocity v_w and slip length b on a plane wall.

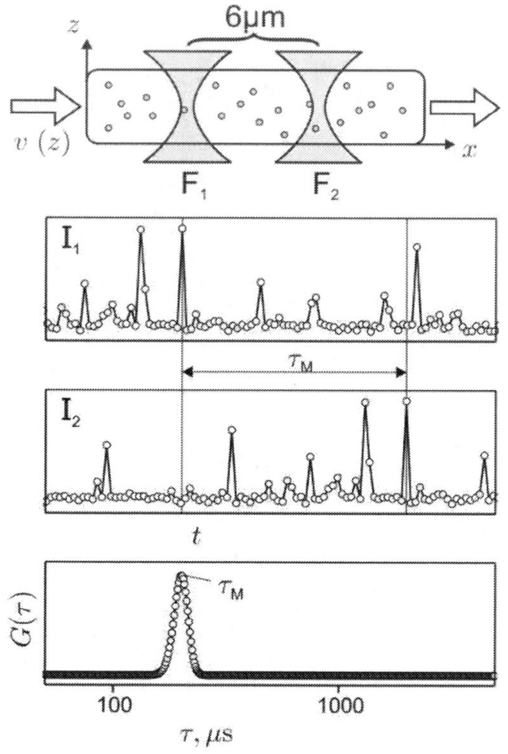

Figure 2. Sketch and principle of the DF-FCS experimental technique (from [2]; ordinates in arbitrary units).

Even though it is now admitted that the no-slip condition usually applies at ordinary scales, this idea of a slip condition reappears in connection with novel applications in microfluidics. For a gas at the scale of the mean free path, it is known that a slip boundary condition applies for the flow outside the Knudsen layer. But it may also apply to a liquid, e.g. for micro-channels with chemically treated surfaces. Obviously, the slip depends on the liquid and wall materials. Fluids considered here are Newtonian. Non-Newtonian fluids, like polymer solutions, require some particular rheological bulk equations and slip boundary conditions. Even for Newtonian liquids, there is a large body of literature and values of b are a subject of discussions. A review of experiments in various systems is given in [6].

There may also be a slip boundary condition on suspended particles in a liquid, either because of their hydrophobic surface or because they are porous, like for a polymer jumble. Indeed, it has been shown by [7] that a slip applies on a porous wall, the slip length being of the order of the pore size. The scope of this paper is restricted to the case of the no-slip boundary condition on particles.

For applications, the advantage of a slip condition is that pushing a fluid through a micro-channel requires a reduced pressure gradient (which otherwise should be quite large). The incidence of a wall slip on dilute suspensions is considered here. This problem is relevant in microfluidics, for instance for separation techniques in analytical chemistry like field-flow-fractionation (FFF) and capillary hydrodynamic chromatography (CHDC) (see [8, 9, 10]). The influence of fluid slip in CHDC was considered recently in [11].

The outline of the paper is as follows. First, recent experiments for measuring the slip length are presented in §2. Important physical parameters which may be responsible for discrepancies in the results are displayed. The importance of hydrodynamic interactions of particles with a slip wall are emphasised. This is an incentive for studying in detail the hydrodynamic interaction of a spherical particle with a slip wall, when in creeping flow, §3. The general case of a moving sphere in a parabolic shear flow is expressed by linearity of Stokes equations of fluid motion as the superposition of elementary flow problems. The solution for an ambient quadratic flow is presented here in §4. It is searched in terms of series in bispherical coordinates and a fast algorithm is then applied to calculate the coefficients in the series. Results for the example case of a fixed sphere embedded in a quadratic shear flow are presented in §5. Results for a freely moving sphere are provided in §6. Then §7 is concerned with Brownian particles entrained in a shear flow near a slip wall. This problem is relevant for the measurements described in §2 and also for the separation techniques briefly mentioned above. Finally, the conclusion and outlooks are in §8.

2. Measurements of slip length and important parameters

The measurements presented here were performed with the double-focus fluorescence cross-correlation technique (DF-FCS). They are fully documented in [1, 2]. The setup and principle of DF-FCS are sketched in Fig. 2.

A laser beam (543 nm line of a 5 mW Helium-Neon laser) is split as two parallel laser beams separated by a distance of $6\,\mu$m. The beams cross the slip wall from below. Their foci F_1 and F_2 provide two spheroidal test volumes located in the fluid region. Fluorescent tracers are carried along the wall in the x direction by the fluid flow with velocity $v(z)$, where z denotes the coordinate normal to the wall. They cross consecutively the two foci, producing two time-resolved fluorescence intensities $I_1(t)$ and $I_2(t)$. The time cross correlation function calculated as $G(\tau) = \langle I_1(t)I_2(t+\tau)\rangle_t/\langle I_1(t)\rangle_t\langle I_2(t)\rangle_t$ has a maximum at some $\tau = \tau_m$ which is characteristic of the transfer time between foci.

The tracers were either 40 nm diameter spherical latex particles or single dye *Alexa 568* dye molecules with a hydrodynamic diameter of ca. 1.2 nm. The liquid was water, with various proportions of added dissolved salt to screen out the repulsive electrical forces between particles. The tracers were observed with a confocal microscope.

Measured velocities are averages over those of particles located in the spheroidal test volumes. The ζ position refers to the position of the center of each spheroidal test volume relative to the channel midplane. A typical velocity profile is shown in Fig. 3. As expected, across the channel (left) the parabolic Poiseuille profile is obtained. The plots (a) and (b) in Fig. 3 (right) show enlargements of the near-wall regions, the walls being represented as dashed lines. Note that ζ may exceed the distance between the channel wall and the channel midplane. When the whole spheroidal test volume is in the fluid, all particles are illuminated and contribute to the signal. When the center of the test volume is in the transparent wall (the glass slide) while a part of this volume is still in the fluid, some particles close to the wall are still illuminated. The signal then is weaker but provides a value of the velocity in the vicinity of the wall. The plateau, in either (a) or (b), then corresponds to particles located very close to a wall. The non-zero value at the plateau exhibits a slip velocity on the wall. From these figures, the slip length may then be obtained, see Fig. 1.

Results for the slip length presented in Tab. 1 show wide variations depending upon the used particles, fluid and wall material. This illustrates the difficulties involved in using particle

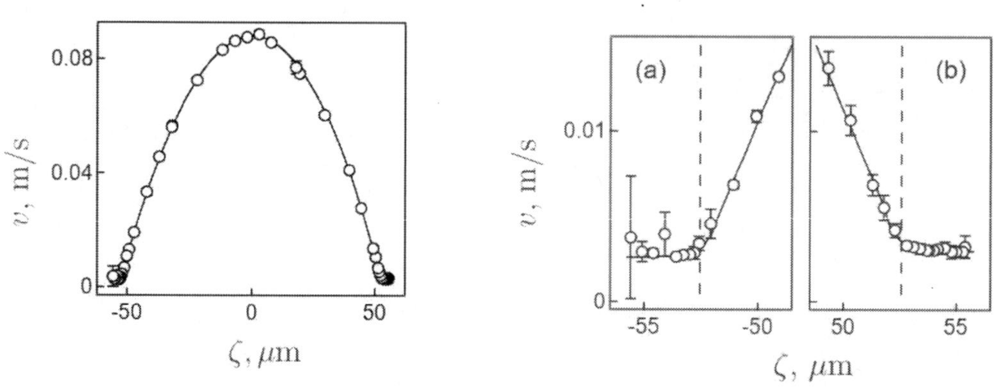

Figure 3. Measured velocity of particles located in the test volume, versus the position ζ of the center of the test volume (from [1]). Particles are latex spheres. Left: across channel. Right: near walls, showing the slip velocity on the walls. The experimentally determined positions of the walls are shown as dashed lines. When the center of test volume is in the solid transparent wall, some particles in the fluid close to the wall provide a value of the velocity there. This velocity appears as a plateau in the walls (a) and (b).

tracking techniques (like this technique, and also micro-PIV) to measure a slip length.

	glass slide ⇒	borosilicate	glass slide	mica
Latex spheres,	salt-free water	1.00	0.89	0.86
Alexa 568,	salt-free water	0.82	0.63	0.51
Alexa 568,	0.01 mM salted water	0.32	0.59	
Alexa 568,	1 mM salted water	0.25	0.22	

Table 1. Experimental results for the slip length (in μm).

The differences in the table may be interpreted in terms of various physical phenomena:

(i) Repulsive electrical forces push particles away from wall. Thus they go faster and this increases the apparent slip length. But some values are still not consistent, comparing latex spheres and the Alexa 568 molecules.

(ii) The Aris-Taylor dispersion provides an apparent slip length.

(iii) The hydrodynamic interactions between particles and slip wall modify the relative velocity between the tracer and fluid.

The important influence of item (ii) was resolved in [2]. Experimental results and the model are compared in Fig. 4. The notation for ζ is like in Fig. 3 and the plateau also corresponds to particles located very close to the wall that is shown as a dashed line. In these results:

(i) A 10^{-4} mol/l NaCl solution was used for small repulsive forces.

(ii) Only particles at a distance from the wall were considered in fitting the model and experiment, so that particle-wall interactions could be neglected.

The calculated Aris-Taylor dispersion with these conditions gave a rather large apparent slip length of 740 nm. But subtracting the apparent slip lengths for slip and no-slip walls could provide a true slip length of around 60 to 70 nm.

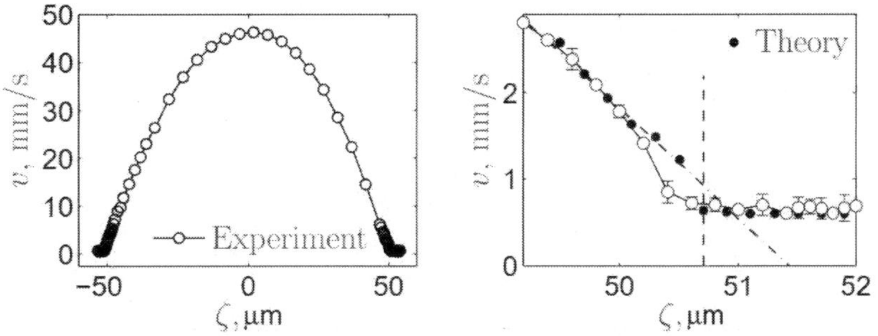

Figure 4. Experimental results using latex spheres and model from [2]. The experimentally determined position of the wall is shown as a dashed line. As for the plateau, see the explanation in the caption of Fig. 3 and in the text.

Yet, as observed in Fig. 4 (right), there is a discrepancy between the theoretical and experimental points near the wall which could be due in particular to particle-wall hydrodynamic interactions. Progress in this domain will thus arise from a study of this effect. This paper brings its contribution in this respect.

3. Hydrodynamical problem for a spherical particle near a slip wall

Considering the size of particles involved in the measurement techniques of the slip length and in separation methods, the Reynolds number for the flow field relative to a particle is usually low as compared with unity and Stokes equations for creeping flow apply. Particles considered here are solid and spherical. The suspension of particles is dilute in the sense that the volume fraction is low. Then, hydrodynamic interactions between particles may be neglected. That is, particles may be considered as independent in their motion relative to the fluid. The typical problem at hand for the quoted application is that of a particle embedded in a fluid that is moving with a given velocity $v(z)$ in the x direction (following the notation of the preceding section) along and in between parallel walls.

There is a large body of literature for spherical particles in creeping flow near no-slip walls (see e.g. [12, 13] and more recent references in [14]), but less results are available for a slip wall.

It is shown in [14] that, in view of modeling the hydrodynamic interactions between a particle and two no-slip walls, the approximation of considering only the hydrodynamic interactions with the nearest no-slip wall has a surprisingly large range of application. Thus, the approximation of considering the nearest slip wall may be sufficient for many purposes. This idea is based on the fact that hydrodynamic interactions with a slip wall are weaker than with a no-slip wall.

The presentation follows this line. Consider (Fig. 5) a spherical particle with radius a centred at a distance ℓ from a slip wall. The particle is moving as a solid body with translational velocity $\mathbf{U}$ and rotational velocity $\mathbf{\Omega}$ in an ambient flow field with velocity $\mathbf{v} = [k_s(z+b) + k_q z^2]\,\mathbf{e}_x$ along the wall. Here, k_s and k_q denote constant coefficients and $\mathbf{e}_x$ is the unit vector along x. The perturbed flow velocity $\mathbf{u}$ and pressure p satisfy the Stokes equations:

$$\nabla \cdot \mathbf{u} = 0 \tag{1a}$$

$$\mu \nabla^2 \mathbf{u} = \nabla p, \tag{1b}$$

with boundary conditions:

$$\mathbf{u} = b\frac{d\mathbf{u}}{dz} \quad \text{on} \quad z = 0 \tag{2a}$$

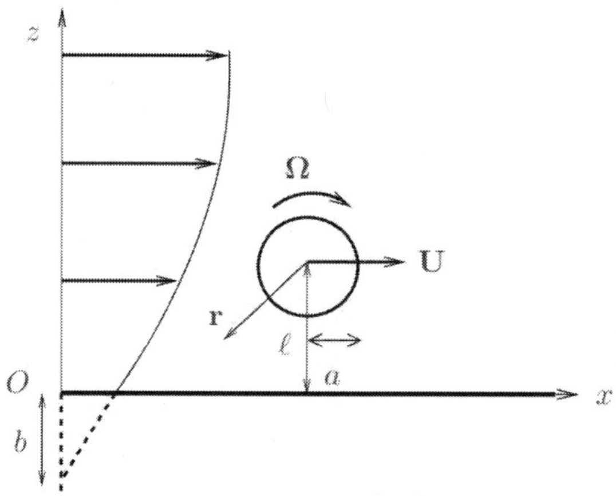

Figure 5. Notation for a solid spherical particle moving in an ambient flow field near a slip wall.

$$\mathbf{u} \to \mathbf{v} = [k_s(z+b) + k_q z^2]\,\mathbf{e}_x \quad \text{for} \quad |\mathbf{r}| \to \infty \tag{2b}$$

$$\mathbf{u} = \mathbf{U} + \boldsymbol{\Omega} \times \mathbf{r} \quad \text{on} \quad |\mathbf{r}| = a, \tag{2c}$$

where $\mathbf{r}$ denotes a position vector originating from the sphere centre. By linearity of the Stokes equations and boundary conditions, this problem may be solved as the superposition of the cases of:

(i) a sphere translating ($\mathbf{U}$) and rotating ($\boldsymbol{\Omega}$) in a fluid at rest,

(ii) a fixed sphere in a linear shear flow ($k_s(z+b)\,\mathbf{e}_x$),

(iii) a fixed sphere in a quadratic shear flow ($k_q z^2\,\mathbf{e}_x$).

Problem (i) was solved by the method of bispherical coordinates in [15]. The system of equations was rearranged in a simpler way in [3]. Reference [3] also treated problem (ii) along the same line. Problems (i) and (ii) were also considered by [16] using the boundary integral method. Problem (iii) presented in [4] is exposed below; the analytical approach follows that of [3].

Quantities of interest in view of modeling the particle motion are the force $\mathbf{F}$ and torque $\mathbf{C}$ on the particle, calculated from:

$$\mathbf{F} = \int_S \boldsymbol{\sigma} \cdot \mathbf{n}\, dS \tag{3a}$$

$$\mathbf{C} = \int_S \mathbf{r} \times \boldsymbol{\sigma} \cdot \mathbf{n}\, dS, \tag{3b}$$

where $\boldsymbol{\sigma}$ is the stress tensor, $\mathbf{n}$ an unit normal vector and S the sphere surface. From the particle equations of motion, the velocities $\mathbf{U}$ and $\boldsymbol{\Omega}$ (which are assumed to be given in the preceding step) can then be obtained, e.g. for a freely moving particle (that is for zero external force and torque).

The symmetric first moment of stresses, or "stresslet":

$$\mathbf{S} = \int_S \left[\frac{1}{2}(\mathbf{r}\boldsymbol{\sigma} \cdot \mathbf{n} + \boldsymbol{\sigma} \cdot \mathbf{n}\mathbf{r}) - \frac{1}{3}\mathbf{I}(\mathbf{r} \cdot \boldsymbol{\sigma} \cdot \mathbf{n})\right] dS \tag{4}$$

where $\mathbf{I}$ is the identity tensor, is also of interest, in view of calculating the rheology of a dilute suspension.

The force, torque and stresslet for problem (1)(2) are obtained from the preceding decomposition as the sum of those for the elementary problems. These quantities may be written in terms of friction factors as:

Translation :	$F_x^t = -6\pi a\mu U f_{xx}^t$,	$C_y^t = 8\pi a^2\mu U c_{yx}^t$,	$S_{xz}^t = 6\pi a^2\mu U s_{xz}^t$
Rotation :	$F_x^r = 6\pi a^2\mu\Omega f_{xy}^r$,	$C_y^r = -8\pi a^3\mu\Omega c_{yy}^r$,	$S_{xz}^r = 6\pi a^3\mu\Omega s_{xz}^r$
Linear shear flow :	$F_x^s = 6\pi a\mu(k_s\ell + b)f_{xx}^s$,	$C_y^s = 4\pi a^3\mu k_s c_{yx}^s$,	$S_{xz}^s = \frac{10}{3}\pi a^3\mu k_s s_{xz}^s$
Quadratic shear flow :	$F_x^q = 6\pi a\ell^2\mu k_q f_{xx}^q$,	$C_y^q = 8\pi a^3\ell\mu k_q c_{yx}^q$,	$S_{xz}^q = \frac{20}{3}\pi a^3\ell\mu k_q s_{xz}^q$

(5)

The force and torque for translation, rotation and linear shear flow (in green in the table) were calculated by [3]. The stresslet for translation and rotation and the force, torque and stresslet for quadratic shear flow (in red in the table) were obtained more recently [4]. Forces, torques and stresslets are here normalised by their counterpart for a sphere in unbounded fluid ($\ell/a \to \infty$), obtained from Faxen's formulae. Then, all normalised quantities, or "friction factors" (denoted with lower case letters, in obvious notation) are dimensionless and become unity for $\ell/a \to \infty$, except for the coupling terms c_{yx}^t (torque due to translation) and f_{xy}^r (force due to rotation) which vanish. There is also a symmetry relationship between these two coefficients from Lorentz reciprocal theorem: $f_{xy}^r = \frac{4}{3}c_{yx}^t$.

4. Spherical particle in quadratic flow near a slip wall; solution in bispherical coordinates

Using cylindrical coordinates (ρ, z, ϕ), see Fig. 6 (the sphere centre being at $\rho = 0, z = \ell$), the perturbation flow field $\{\mathbf{w} = \mathbf{v} - \mathbf{u},\ q = p - 2\mu k_q\}$ satisfying the momentum equation (1a) is written following [17, 18] as:

$$w_\rho = \tfrac{1}{2}\{\rho Q_1 + U_0 + U_2\}\cos\phi, \qquad w_z = \frac{1}{2}\{zQ_1 + 2U_1\}\cos\phi, \tag{6a}$$

$$w_\phi = \tfrac{1}{2}(U_2 - U_0)\sin\phi, \qquad q = \mu Q_1 \cos\phi, \tag{6b}$$

where U_0, U_1, U_2, Q_1 are harmonic functions to be found.

These functions are searched for as series in bispherical coordinates. These coordinates have been used in a number of problems for nearly 100 years. Definition formulae of the bispherical coordinates (η, ξ, ϕ) and a sketch are displayed in Fig. 6. More details are in [12], Appendix A-19.

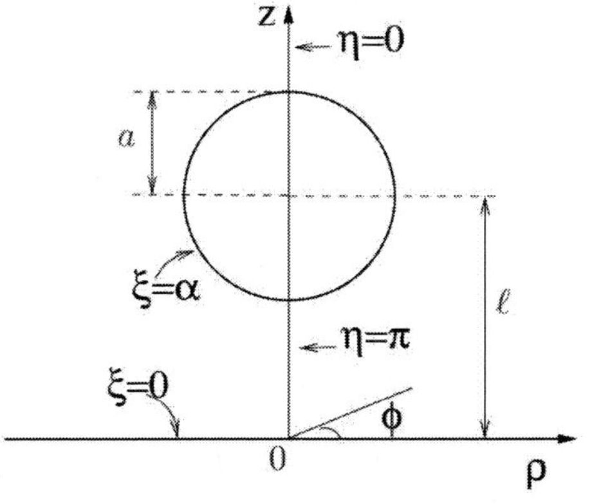

$$\rho = d\frac{\sin\eta}{\cosh\xi - \cos\eta}$$

$$z = d\frac{\sinh\xi}{\cosh\xi - \cos\eta}$$

$$d = \sqrt{\ell^2 - a^2}$$

$$(\eta, \xi, \phi) \in [0, \pi] \times [0, \alpha] \times [0, 2\pi]$$

Figure 6. Definition of bispherical coordinates (η, ξ, ϕ).

Note that $\xi = 0$ represents the plane $z = 0$ and $\xi = \alpha$ represents the sphere surface, with $\ell = a\cosh\alpha$. The application of boundary conditions on the plane and sphere is then made easier.

The expressions of the harmonic functions U_1, Q_1, U_0, U_2 as series in these coordinates are:

$$U_1 = d^2(\cosh\xi - \beta)^{1/2}\sin\eta\sum_{n\geq 1} A_n\sinh(\gamma_n\xi)P_n'(\beta), \tag{7a}$$

$$Q_1 = d(\cosh\xi - \beta)^{1/2}\sin\eta\sum_{n\geq 1}[B_n\cosh(\gamma_n\xi) + C_n\sinh(\gamma_n\xi)]P_n'(\beta), \tag{7b}$$

$$U_0 = d^2(\cosh\xi - \beta)^{1/2}\sum_{n\geq 0}[D_n\cosh(\gamma_n\xi) + E_n\sinh(\gamma_n\xi)]P_n(\beta), \tag{7c}$$

$$U_2 = d^2(\cosh\xi - \beta)^{1/2}\sin^2\eta\sum_{n\geq 2}[F_n\cosh(\gamma_n\xi) + G_n\sinh(\gamma_n\xi)]P_n''(\beta), \tag{7d}$$

where $\beta = \cos\eta$, $\gamma_n = n + 1/2$, $P_n(\beta)$ is the Legendre polynomial of order n and $(\ )' = \partial/\partial\beta$. By construction, the harmonic functions vanish at infinity which is represented by $\xi = \eta = 0$ in bispherical coordinates.

The coefficients $A_n, \cdots, G_n$ in the series (7) have to be found from the boundary conditions and also from the continuity equation (1a). Calculations are made simpler by using the following formula, obtained by differentiating the slip boundary condition (2a) and combining with the continuity equation (1a):

$$\frac{\partial w_z}{\partial z} = b\frac{\partial^2 w_z}{\partial z^2} \quad \text{on} \quad z = 0. \tag{8}$$

As a result, the unknown coefficients B_n, D_n, F_n are obtained in terms of the other ones A_n, C_n, E_n, G_n. The system to be solved for A_n, C_n, E_n, G_n may then be written in the form:

$$\hat{M}_1 X_1 + \hat{M}_1^+ X_2 = \mathcal{B}_1 \tag{9a}$$

$$M_n^- X_{n-1} + M_n X_n + M_n^+ X_{n+1} = \mathcal{B}_n \qquad (n \geq 2) \tag{9b}$$

where the M's are 4×4 known matrices, the $\mathcal{B}$'s are known vectors and the unknowns are:

$$X_1 = \begin{pmatrix} A_1 \\ C_1 \\ E_1 \\ E_0 \end{pmatrix}; \qquad X_n = \begin{pmatrix} A_n \\ C_n \\ E_n \\ G_n \end{pmatrix} \qquad (n \geq 2)$$

The forms of X_1 and of the matrices $\hat{M}_1$, $\hat{M}_1^+$ and $\mathcal{B}_1$ in (9a) are special because the series in (7) start at different values of n. Relationships between coefficients were then grouped for conciseness. Details [19] are too lengthy to be reproduced here. The matrices M are, perforce, the same as for the linear shear flow [3]. Only the $\mathcal{B}$'s are different, viz.:

$$\mathcal{B}_1 = \begin{pmatrix} 0 \\ K_0 \\ K_1 \\ 0 \end{pmatrix}; \qquad \mathcal{B}_n = \begin{pmatrix} 0 \\ 0 \\ K_n \\ 0 \end{pmatrix} \qquad (n \geq 2)$$

with

$$K_n = \frac{4\sqrt{2}}{3}\frac{(2n+1)(2n+1+\frac{2}{\tanh\alpha})}{1 + e^{(2n+1)\alpha}}. \tag{10}$$

The infinite linear system (9) is closed by noting that all coefficients $A_n, \cdots, G_n$ should vanish for $n \to \infty$ in order that the series in (7) converge. Then,

$$X_n \to 0 \text{ for } n \to \infty. \tag{11}$$

The infinite linear system (9) is solved by truncating for some large n, say N, that is $X_{N+1} = 0$. This gives from (9b) the equation for $n = N$:

$$M_N^- X_{N-1} + M_N X_N = \mathcal{B}_N. \tag{12}$$

A classic way to solve the system $\mathcal{L}$={(9a), (9b) for $2 \le n \le N-1$, (12)} is first to rearrange it as a linear system for the $4N$ unknowns $(A_1, C_1, E_1, E_0, A_2, C_2, E_2, G_2, \cdots, A_N, C_N, E_N, G_N)$. By construction, the system has a banded matrix. The system is solved by LU decomposition. At the next step, N is increased by unity, the novel linear system is solved again, etc., until the results for A_1, C_1, E_0, G_2 converge. It is then a simple matter to derive all coefficients by the recurrence relationships (9). The method amounts to solving $4N \times 4N$ linear systems of increasingly larger sizes, which usually requires a large amount of computer calculation time and memory. Moreover, when $\ell/a - 1$ is moderate, obtaining a number of terms in the expansions is necessary in order to have a sufficient precision.

A faster method for solving an infinite linear system like (9), but for scalars, was proposed by [20] and applied thereafter to solve other hydrodynamic problems involving a sphere and a plane in [21, 22]. One could think of extending that method to matrices. However, this is not possible here, since the matrices $\hat{M}_1^+$ and M_n^+ $(n \ge 2)$ in (9) have zero determinant and are thus not invertible.

For these reasons, the following faster algorithm is proposed. The algorithm for inverting the truncated block tridiagonal matrix $\mathcal{L}$ is basically an extension of Thomas' algorithm for solving a tridiagonal system. It proceeds as follows:

- The formal solution of (9a) for X_1 is $X_1 = U_1 X_2 + V_1$ with:

$$U_1 = -[\hat{M}_1]^{-1}\hat{M}_1^+; \qquad V_1 = [\hat{M}_1]^{-1}\mathcal{B}_1. \tag{13}$$

- The formal solution of (9b) for X_n when $2 \ge n \ge N-1$ is $X_n = U_n X_{n+1} + V_n$ with:

$$U_n = -[M_n^- U_{n-1} + M_n]^{-1} M_n^+; \qquad V_n = [M_n^- U_{n-1} + M_n]^{-1}[B_n - M_n^- V_{n-1}]. \tag{14}$$

- The formal solution of (12) for X_N is $X_N = V_N$ with:

$$V_N = [M_N^- U_{N-1} + M_N]^{-1}[B_N - M_N^- V_{N-1}]. \tag{15}$$

It appears that (U_1, V_1) may be directly calculated from (13), then in turn (U_n, V_n) from (14) for $n = 2, \cdots, N-1$, then V_N from (15). Once V_N is known, $X_N = V_N$, then all other X_n for $n = N-1, \cdots, 1$ are calculated by proceeding downwards.

This procedure has two main advantages:

- At each step, only one 4×4 matrix is inverted, namely $[M_n^- U_{n-1} + M_n]$.
- When N is increased by unity, only U_{N-1} and U_N, V_N should be calculated, thus only the supplementary $[M_N^- U_{N-1} + M_N]$ matrix has to be inverted.
  Then the new X_N and X_n's can be obtained.

This algorithm was implemented in Mathematica© computer algebra language to calculate asymptotic expansions for $a/\ell \ll 1$ (large distance from wall).

5. Results for a spherical particle in quadratic shear flow

By performing the integrations in (3) and (4), the force, torque and stresslet friction factors for a fixed sphere in a quadratic shear flow are obtained in terms of the coefficients of the series (7) as:

$$f^q_{xx} = -\frac{\sqrt{2}}{6}\frac{\sinh^3\alpha}{\cosh^2\alpha}\sum_{n\geq 0}\{E_n + n(n+1)C_n\}, \tag{16a}$$

$$c^q_{yx} = \frac{\sqrt{2}}{4}\frac{\sinh^4\alpha}{\cosh\alpha}\sum_{n\geq 0}\left\{\frac{E_n}{\tanh\alpha} + 2n(n+1)A_n\right\}, \tag{16b}$$

$$s^q_{xz} = \frac{3\sqrt{2}}{20}\frac{\sinh^4\alpha}{\cosh\alpha}\sum_{n\geq 0}\left\{\frac{E_n}{\tanh\alpha} - \frac{1}{3}n(n+1)(2n+1)(B_n + C_n)\right\}. \tag{16c}$$

These friction factors were calculated in terms of the normalised sphere centre to wall gap, $\ell/a-1$ and of the normalised slip length $\lambda = b/a$. Results are displayed in figures 7, 8 and 9[1]. Values of f^q_{xx} and c^q_{yx} for $\lambda = 0$ are in good agreement with those obtained for a sphere in a quadratic flow near a no-slip wall by [22]. As found in [22], values of c^q_{yx} have a small minimum around $\ell/a - 1 \simeq 0.18$, which is not so apparent at this scale. When λ increases, it is observed in Fig. 7 that f^q_{xx} decays. Fig. 8 shows that c^q_{yx} also decays for increasing λ for a dimensionless gap $\ell/a-1$ smaller than around 0.2 but increases otherwise with λ. The minimum at $\lambda = 0, \ell/a - 1 \simeq 0.18$ also disappears and is replaced by a maximum at some larger value of $\ell/a - 1$ for values of λ of order unity. The physical interpretation of these behaviours is not straightforward, since it depends on a complicated flow structure, see e.g. Fig. 3 of [22] for the case $\lambda = 0$.

Asymptotic expansions were obtained for large sphere to wall distances $a/\ell \ll 1$; for example, for the force friction factor:

$$\begin{aligned} f^q_{xx} = {} & 1 + \frac{9}{16}\left(\frac{a}{\ell}\right) + \left(\frac{499}{768} - \frac{9\lambda}{16}\right)\left(\frac{a}{\ell}\right)^2 + \left(\frac{9\lambda^2}{8} - \frac{81\lambda}{128} - \frac{1575}{4096}\right)\left(\frac{a}{\ell}\right)^3 \\ & + \left(-\frac{81\lambda^3}{16} + \frac{405\lambda^2}{256} + \frac{3701\lambda}{4096} - \frac{7263}{65536}\right)\left(\frac{a}{\ell}\right)^4 \\ & + \left(\frac{261\lambda^4}{8} - \frac{891\lambda^3}{128} - \frac{19551\lambda^2}{4096} + \frac{4959\lambda}{16384} + \frac{1596155}{3145728}\right)\left(\frac{a}{\ell}\right)^5 \qquad (17a) \\ & + \left(-\frac{4005\lambda^5}{16} + \frac{5589\lambda^4}{128} + \frac{17507\lambda^3}{512} - \frac{45081\lambda^2}{32768} - \frac{2197069\lambda}{1048576} - \frac{3036175}{16777216}\right)\left(\frac{a}{\ell}\right)^6 \\ & + \ldots \qquad (17b) \end{aligned}$$

Up to 14 terms could be calculated in 5 mn on a PC using *Mathematica* computer algebra software. Similar results for the torque and stresslet are provided in [4]. Such asymptotic expansions are by definition valid provided each term is of lower order than the preceding one, giving the supplementary condition for validity: $\lambda a/\ell \ll 1$. Values calculated with these expansions are compared to the accurate results in figures 7, 8 and 9.

6. A freely moving sphere near a slip wall

Based upon the decomposition in §3, earlier results and the results of §4, the case of a freely moving sphere in the ambient flow field $\mathbf{v} = [k_s(z+b) + k_q z^2]\,\mathbf{e}_x$ can now be considered. Writing that the net force and torque vanish:

$$F^t_x + F^r_x + F^s_x + F^q_x = 0, \qquad C^t_y + C^r_y + C^s_y + C^q_y = 0 \tag{18}$$

[1] Figures in this section are reprinted with permission from [4], ©2011, American Institute of Physics.

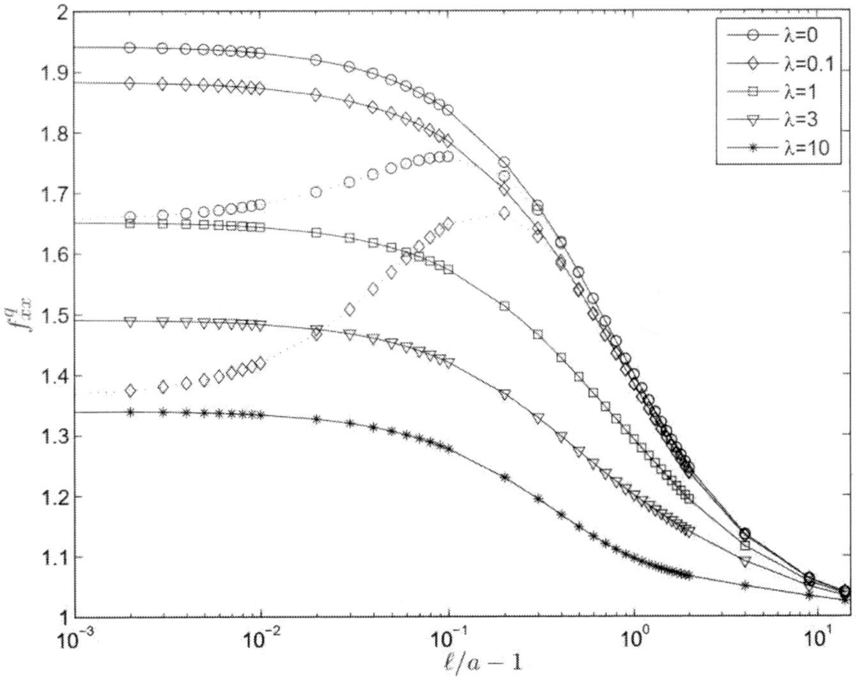

Figure 7. Friction factor for the force on a fixed sphere in a quadratic flow. Solid lines: accurate results. Dotted lines: asymptotic results for $a/\ell \ll 1$.

provides the translational velocity $\mathbf{U} = U\mathbf{e}_x$ and rotational velocity $\mathbf{\Omega} = \Omega\mathbf{e}_y$ of the sphere with:

$$U = U_s + U_q, \qquad \Omega = \Omega_s + \Omega_q, \tag{19}$$

in which reduced values of U_s, Ω_s for the linear shear flow were derived in [3] to be:

$$u_s = \frac{U_s}{k_s(\ell+b)} = \frac{\left(\frac{\ell}{a}\right) c^r_{yy} f^s_{xx} + \frac{1}{2} f^r_{xy} c^s_{yx}}{\left(\frac{\ell}{a}+\lambda\right)\left(c^r_{yy} f^t_{xx} - f^r_{xy} c^t_{yx}\right)}, \tag{20a}$$

$$\omega_s = \frac{\Omega_s}{k_s/2} = \frac{c^s_{yx} f^t_{xx} + 2\left(\frac{\ell}{a}\right) f^s_{xx} c^t_{yx}}{c^r_{yy} f^t_{xx} - f^r_{xy} c^t_{yx}} \tag{20b}$$

and reduced values of U_q, Ω_q for the quadratic shear flow are [4]:

$$u_q = \frac{U_q}{k_q\left(\ell^2 + \frac{a^2}{3}\right)} = \frac{\left(\frac{\ell}{a}\right)^2 c^r_{yy} f^q_{xx} + \left(\frac{\ell}{a}\right) f^r_{xy} c^q_{yx}}{\left(\left(\frac{\ell}{a}\right)^2 + \frac{1}{3}\right)\left(c^r_{yy} f^t_{xx} - f^r_{xy} c^t_{yx}\right)}, \tag{21a}$$

$$\omega_q = \frac{\Omega_q}{k_q \ell} = \frac{c^q_{yx} f^t_{xx} + \left(\frac{\ell}{a}\right) f^q_{xx} c^t_{yx}}{c^r_{yy} f^t_{xx} - f^r_{xy} c^t_{yx}}. \tag{21b}$$

Values of the reduced translational velocity are plotted in Fig. 10[2]. The velocity increase with λ is due to the reduced drag on the wall when the slip increases. The velocity can be even larger than that far from the wall when λ is large. This can be understood since, roughly speaking,

[2] Figures in this section are reprinted with permission from [4], ©2011, American Institute of Physics.

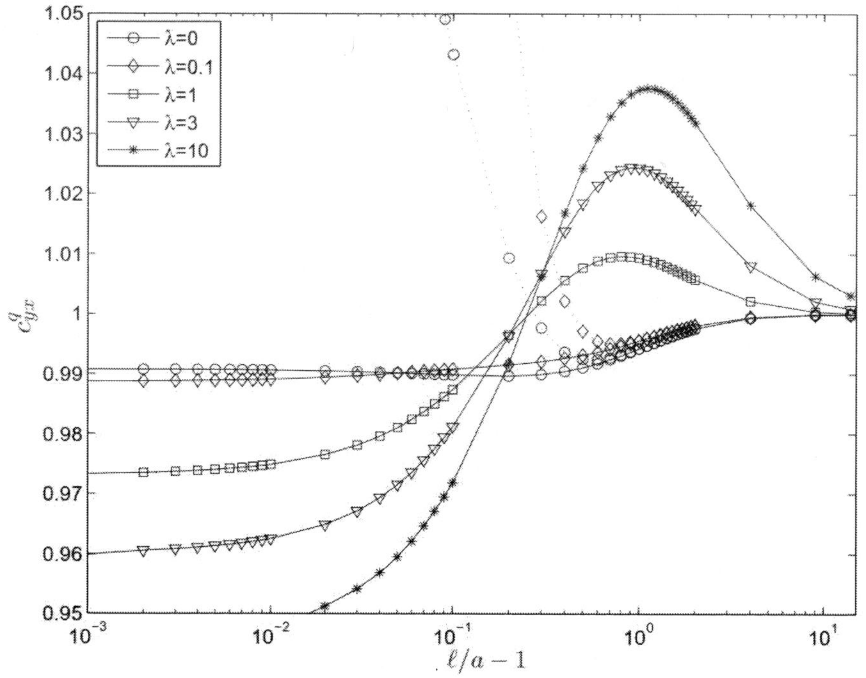

Figure 8. Friction factor for the torque on a fixed sphere in a quadratic flow. Solid lines: accurate results. Dotted lines: asymptotic results for $a/\ell \ll 1$.

the particle close to the wall is then submitted to viscous stresses only on one side of its surface. Based upon the series expansions for large ℓ/a, a Padé approximant was constructed for u_q. The results it provides (see dashed lines in Fig. 10) are in good agreement with the accurate results, even down to small gaps of around $\ell/a - 1 \simeq 0.2$. Values of the reduced rotational velocity are plotted in Fig. 11. It is seen that this quantity has small variations with λ.

The xz component of the stresslet for a freely moving sphere in the ambient flow field $\mathbf{v} = [k_s(z+b) + k_q z^2]\,\mathbf{e}_x$ can also by linearity be expanded as:

$$S_{xz} = S^t_{xz} + S^r_{xz} + S^s_{xz} + S^q_{xz}. \qquad (22)$$

The normalised stresslet for the freely moving sphere in quadratic flow field is then obtained as [4]:

$$s^{qf}_{xz} = \frac{S^{qf}_{xz}}{\frac{20}{3}\pi a^3 \ell \mu k_q} = \left(\frac{9}{10}\frac{\ell}{a} + \frac{3}{10}\frac{a}{\ell}\right) s^t_{xz} u_q + \frac{9}{10} s^r_{xz}\omega_q + s^q_{xz}, \qquad (23)$$

with the normalised u_q, ω_q, see Eq. (21), and the reduced stresslets:

$$s^t_{xz} = \frac{S^t_{xz}}{6\pi a^2 \mu U}, \qquad s^r_{xz} = \frac{S^r_{xz}}{6\pi a^3 \mu \Omega}, \qquad s^q_{xz} = \frac{S^q_{xz}}{\frac{20}{3}\pi a^3 \ell \mu k_q}, \qquad (24)$$

which have the limit of unity when $\ell/a \to \infty$.

The reduced stresslet s^{qf}_{xz} is plotted in Fig. 12. Padé approximants are seen to fit quite well the accurate results, even close the wall. The stresslet is relevant for the effective viscosity of a dilute suspension near a slip wall. A lower stresslet for larger slip suggests that the effective viscosity of a suspension would be smaller for slip walls.

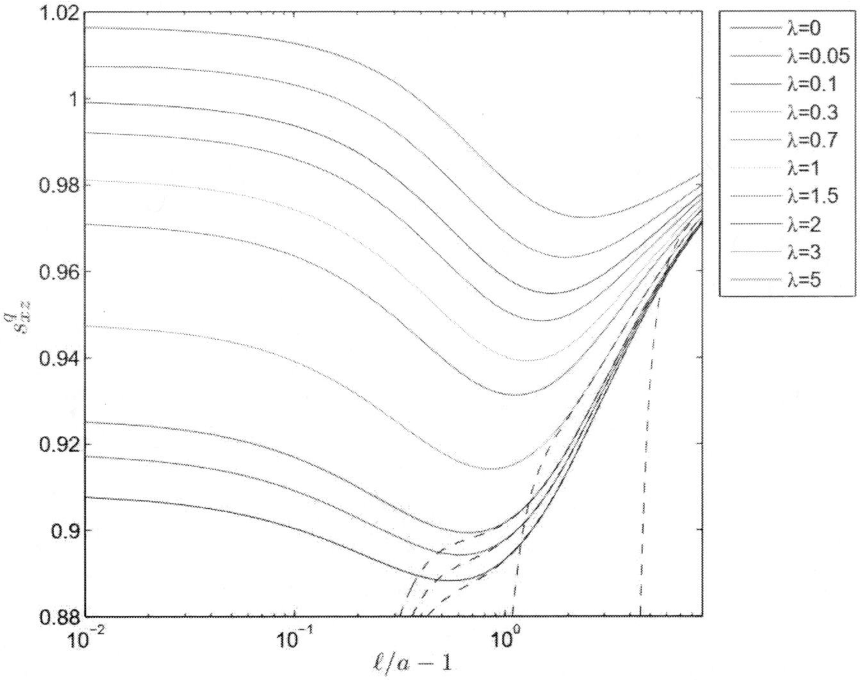

Figure 9. Friction factor for the stresslet on a fixed sphere in a quadratic flow. Values of the reduced slip length λ correspond to curves from bottom to top. Solid lines: accurate results. Dashed lines: asymptotic results for $a/\ell \ll 1$.

7. Aris-Taylor dispersion of suspension near slip wall

The motion of Brownian particles entrained in a shear flow near a slip wall is relevant for various applications, the measurement technique mentioned in §2 and also separation techniques in analytical chemistry.

When Brownian motion combines with entrainment by an ambient shear flow, Aris-Taylor dispersion of particles occurs. The principle is that:

- Particles diffuse by Brownian motion in z direction normal to the wall to a position where they are entrained by the shear flow along the wall.
- The larger their distance to the wall, the larger their velocity along it.
- They are then widely dispersed in the x direction.

To model this phenomenon, we consider a dilute suspension of equal spherical particles. We write an advection-diffusion equation for the concentration $c(x, y, z, t)$ in particles (any type of definition of the concentration can be used; here, we use e.g. the number of particles per unit volume):

$$\frac{\partial c}{\partial t} + \frac{\partial [U_x c]}{\partial x} + \frac{\partial [U_z c]}{\partial z} = \nabla \cdot [\mathbf{D} \cdot \nabla c], \tag{25}$$

in which U_x, U_z are the components of the particles at some position (x, y, z) and some time t. The motion U_x along x is essentially due to advection by the shear flow, whereas U_z contains the effect of migration due to electrostatic repulsive forces from the wall. $\mathbf{D}$ denotes the Brownian

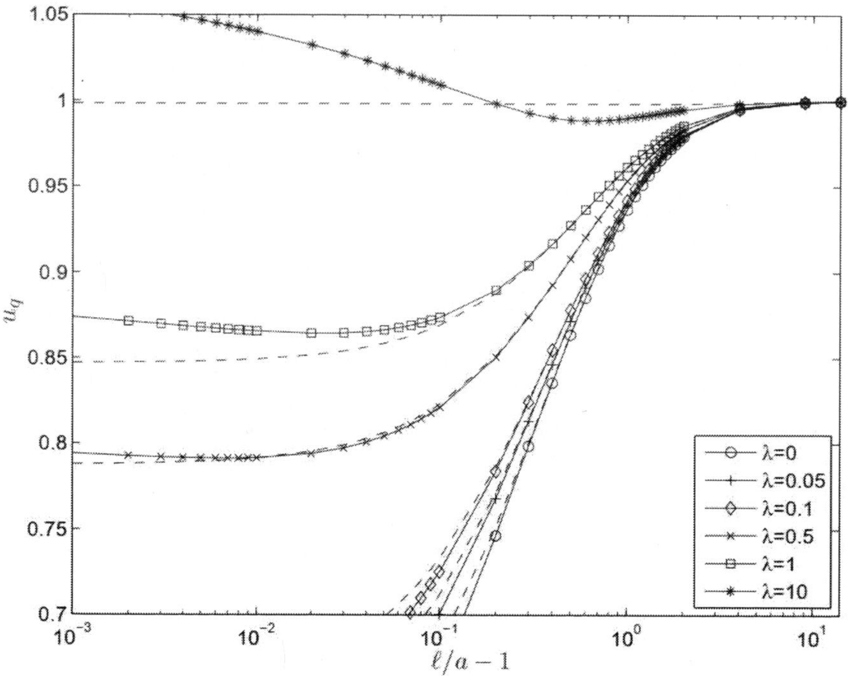

Figure 10. Normalised translational velocity of a freely moving sphere in a quadratic flow near a slip wall. Solid lines: accurate results. Dashed lines: Padé approximants for small a/ℓ.

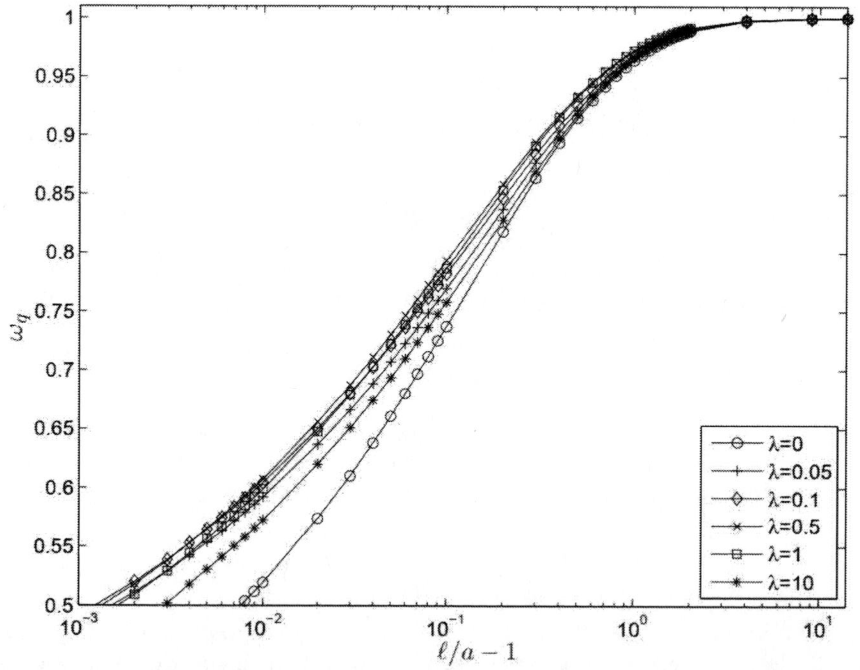

Figure 11. Normalised rotational velocity of a freely moving sphere in a quadratic flow near a slip wall. Accurate results.

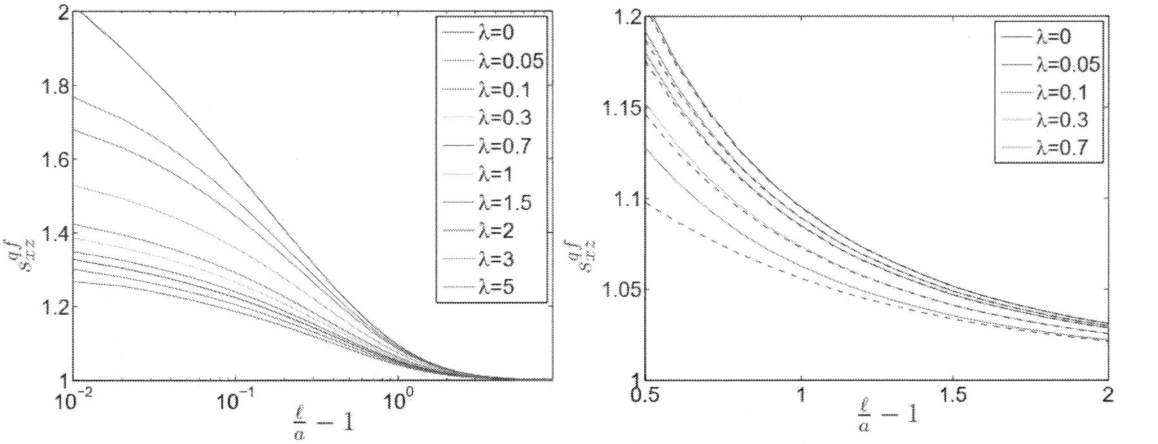

Figure 12. Stresslet for a freely moving sphere in quadratic shear flow near a slip wall. Solid lines: accurate results. Dashed lines (right in zoomed figure): Padé approximants for small a/ℓ.

diffusion tensor:

$$\mathbf{D} = \begin{pmatrix} D_x & 0 & 0 \\ 0 & D_y & 0 \\ 0 & 0 & D_z \end{pmatrix}. \tag{26}$$

The components depend on the distance to the wall z and on the slip length b. Note that $D_y = D_x$ by symmetry.

The Brownian diffusion tensor (26) is diagonal for spherical particles. Indeed, using Einstein's argument, Brownian motion is equivalent to a thermodynamic force, that is the gradient of the chemical potential (Helmholtz free energy per unit mole). Then, following arguments of [23], a spherical particle close to a wall and submitted to an external force that is either normal or parallel to the wall moves in the same direction as this force. This result comes from symmetry of the figure and linearity of Stokes equations of fluid motion.

More precisely, the diffusion coefficient D_x for freely moving spheres along a slip wall may be obtained from Einstein's argument, which is repeated as follows. The thermodynamic force for a dilute suspension is: $\mathbf{F}^B = -kT\boldsymbol{\nabla}\log c$, where k is Boltzmann constant and T the absolute temperature. Brownian motion gives zero torque on a sphere. The balances of forces along x and torques along y are:

$$F_x^t + F_x^r - kT\frac{d}{dx}\log c = 0, \qquad C_y^t + C_y^r = 0. \tag{27}$$

Expressing the forces and torques in terms of velocities and friction coefficients, we obtain in particular the translation velocity U_x along x, and then the flux in the form:

$$U_x c = -D_x \frac{dc}{dx}. \tag{28}$$

This is Fick's law for diffusion, with diffusion coefficient in x direction:

$$D_x = \frac{c_{yy}^r}{f_{xx}^t c_{yy}^r - c_{yx}^t f_{xy}^r} D_0, \tag{29}$$

where $D_0 = \dfrac{kT}{6\pi a\mu}$ is Einstein's diffusion coefficient.

Along z normal to wall, there is no rotational velocity by symmetry, so that the balance of forces is simply:

$$F_z^t - kT\frac{d}{dz}\log c = 0, \tag{30}$$

giving:

$$U_z c = -D_z\frac{dc}{dx}. \tag{31}$$

This is again Fick's law, with diffusion coefficient in z direction:

$$D_z = \frac{1}{f_{zz}^t}D_0. \tag{32}$$

The friction coefficient f_{zz}^t for a sphere translating normally to a slip wall may be obtained from the calculations of [20] (after correcting misprints).

The advection-diffusion equation (25) with diffusion tensor given by (26) (29) (32) provides a model for Aris-Taylor dispersion. We apply it now to the DF-FCS experimental technique presented in §2.

First, in a fluid at rest, there is an equilibrium concentration profile $c_0(z)$ of particles submitted on one hand to repulsive electrostatic forces from the wall and on the other hand to Brownian diffusion. Let:

$$i_0(x, y, z) = i_{norm}\exp\left[-2(x^2 + y^2)/r_0^2 - 2(z - H)^2/z_0^2\right] \tag{33}$$

be the 3D profile of the intensity of light at the upstream focus. Here, H denotes the distance of the centre of either focus to the wall, see Fig. 2. The concentration of illuminated particles at the upstream focus is: $c(0, x, y, z) = c_0(x, y, z)i_0(x, y, z)$. This is the initial condition. The calculation proceeds by integrating (25) to follow in time this cloud of illuminated particles. Particles move to the downstream focus (1) where they are then illuminated by the intensity:

$$i_1(x, y, z) = i_{norm}\exp\left[-2((x^2 - s) + y^2)/r_0^2 - 2(z - H)^2/z_0^2\right]. \tag{34}$$

We calculate the total intensity of illuminated particles in the fluid domain $z > 0$:

$$I(t) = \int_{z>0} c(t, x, y, z)i_1(x, y, z)\,dx\,dy\,dz. \tag{35}$$

The time evolution is like in Fig. 2, bottom curve. $I(t)$ goes through a maximum for some transit time.

The numerical integration of (25) was performed with the Comsol© finite elements software. Since the problem is three-dimensional, the mesh could not be made small enough to capture with sufficient precision the maximum of $I(t)$. For this reason, it was found more efficient to use a Fourier-transform of Eq. (25) along the y direction (along the wall and normal to the flow direction). The definition of the Fourier transform used here is:

$$\hat{f}(\eta) = \int_{-\infty}^{+\infty}\exp(-i\eta y)f(y)dy. \tag{36}$$

The transformed Eq. (25) reads:

$$\frac{\partial\hat{c}}{\partial t} + \frac{\partial[U_x\hat{c}]}{\partial x} + \frac{\partial[U_z\hat{c}]}{\partial z} = D_x\frac{\partial^2\hat{c}}{\partial x^2} + \frac{\partial}{\partial z}\left[D_z\frac{\partial\hat{c}}{\partial z}\right] - \eta^2 D_y\hat{c}. \tag{37}$$

The required intensity is then obtained from the convolution theorem (and using symmetry in y) as:

$$I(t) = 2\int_{\eta=0}^{+\infty}\left[\int_{x=-\infty}^{+\infty}\int_{z>0}\hat{c}(t,x,\eta,z)\hat{i}_1(x,\eta,z)dzdx\right]d\eta.$$

Gauss-Legendre algorithm was used for the integration in η. Thus, the two-dimensional equation (37) had to be integrated only for a few values of η. A typical view of the cloud evolution is shown in Fig. 13. The particle cloud is elongated in the x direction due to Aris-Taylor dispersion and the concentration decays due to diffusion. The light intensity at the downstream focus is represented as an ellipsoid in the right-hand-side of the figure. Roughly speaking, the intersection of both spots corresponds to the intensity $I(t)$ that we are considering. A typical variation of $I(t)$ (for [NaCl]$= 10^{-4}$ mol/l, $H = 0$) is shown in Fig. 14 (Note that at the time this calculation was done [19, 3], we did not have yet results for hydrodynamic interactions in quadratic shear flow; thus only interactions in a pure shear flow are considered here). The maximum corresponds to the transit time τ_M between foci and the velocity is taken as the distance between foci ($6\,\mu$m) divided by this transit time.

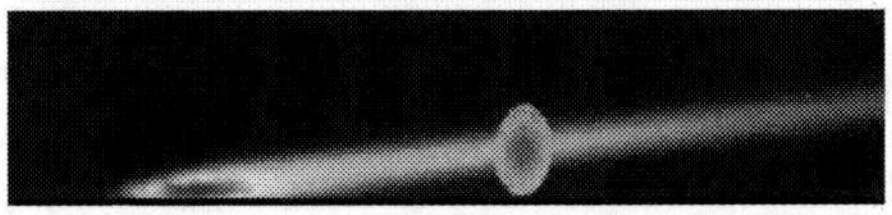

Figure 13. View of the calculated cloud evolution due to Aris-Taylor dispersion and of the light intensity at the downstream focus (from [2]).

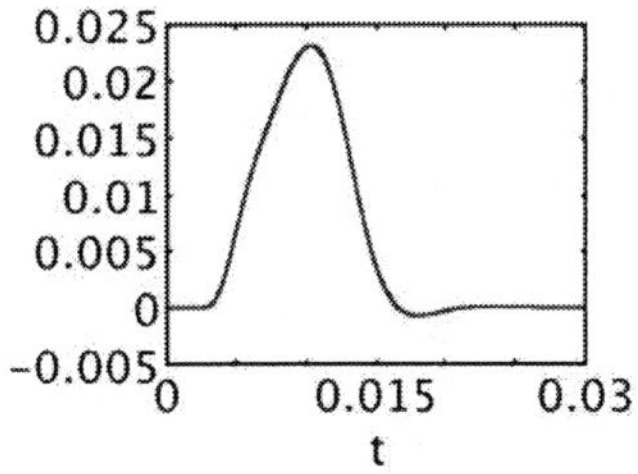

Figure 14. Evolution the calculated light intensity $I(t)$ showing the maximum corresponding to the transit time between foci (from [2]).

By recalculating the cloud evolution for various values of H, we then obtain the velocity profile represented in Fig. 15.

The earlier theoretical results in [2] (see Fig. 4, right) were obtained by neglecting the hydrodynamic interactions with the wall. From the velocity profile, we then obtained a large apparent slip length of 740 nm due to Aris-Taylor dispersion.

If we now compare the theoretical results without hydrodynamic interactions in Fig. 4 to those with interactions in Fig. 15, we observe that the decay close to the wall observed in the experimental results of Fig. 4 (right) is captured in the later case. More calculations have to be done, but these results are encouraging for applying the modeling of near wall hydrodynamic interactions to experimental techniques.

8. Conclusion and outlooks

Experiments [1] show typical difficulties involved in measuring a slip length by following particles. To model this problem, we consider a solid spherical particle moving in an ambient parabolic flow field near a slip wall.

The Stokes flow problem can be decomposed as the sum of elementary ones (translation, rotation of a particle in a fluid at rest and applied linear and quadratic shear flows onto a fixed particle), each of which has a solution as series in bispherical coordinates. The coefficients in the series are solution of 4 coupled recurrence relationships forming an infinite linear system. This

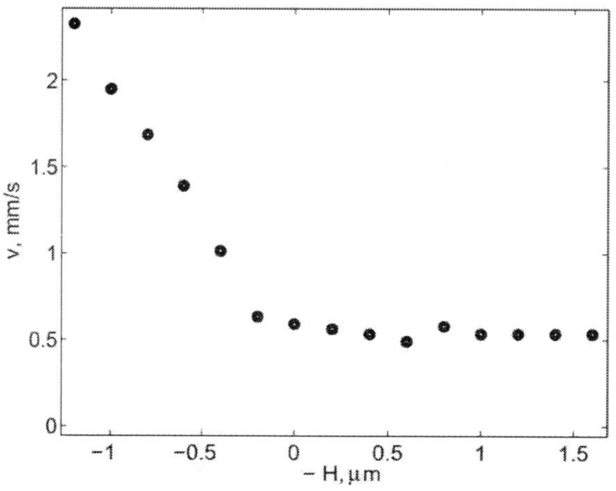

Figure 15. Calculated velocity of particles located in the test volume. H denotes the position of the centre of the test volume with respect to the slip wall. The abscissa $-H$ is used here for comparison with Fig. 4, right. Hydrodynamic interactions when in linear shear flow are used in the calculation. Physical data are the same as in Fig. 4, right.

system is solved by an extension of Thomas' algorithm for inverting a tridiagonal matrix. This algorithm has a cost proportional to the number of terms in the series. It is thus possible to obtain a large number of terms, which are necessary to resolve the lubrication situation when a sphere is close to the slip wall. Accurate results for the force and torque on a sphere, velocity of a freely moving sphere, diffusion tensor are then obtained. The stresslet, that is the symmetric moment of surface stresses on the sphere, is also derived in view of applications to suspension rheology. Expansions of various quantities for a large distance from the wall and for a low slip length are performed on the basis of the analytical solutions, using computer algebra. Padé approximants calculated therefrom provide good approximations.

The Aris-Taylor dispersion of Brownian particles in a shear flow near a slip wall gives a large bias in the measurement of slip [2]. It is calculated from the advection-diffusion equation, using the expressions for the particle velocity and diffusion tensor. It is shown here how wall hydrodynamic interactions with a slip wall modify these results. There are various applications of these results in measurement techniques involving particle tracking and in separation techniques in analytical chemistry.

On the theoretical side, the algorithm for solving this type of infinite linear system can be applied to other problems using the bispherical coordinates technique (in particular asymmetric flow fields problems) so as to obtain results for small gap between surfaces.

The stresslet on a freely moving sphere can be used for deriving the effective viscosity of a dilute suspension of spheres near a slip wall in a linear and parabolic shear flow. Effective collective slip in suspension may then combine with local fluid slip.

The coupled influence of two parallel walls may be important for separation techniques. An extension of the approach of [14] to slip walls may be useful for applications.

As for non-spherical particles near a slip wall, the hydrodynamic problem may be solved by the semi-analytical boundary integral technique. A paper was presented in this symposium on this topic [24].

A non-uniform slip length may also be of interest for applications, e.g. to mixing. A texture optimisation was obtained for a pure fluid flowing between two parallel walls in [25, 26, 27]. Performing optimisations for suspensions would be of interest for microfluidics and mixing processes.

References

[1] Lumma D, Best A, Gansen A, Feuillebois F, Rädler J and Vinogradova O 2003 *Phys. Rev. E* **67** 056313
[2] Vinogradova O I, Koynov K, Best A and Feuillebois F 2009 *Phys. Rev. Lett.* **102** 118302
[3] Feuillebois F, Loussaief H and Pasol L 2009 *CP1186, Applications of Mathematics in Technical and Natural Sciences* ed Todorov M D and Christov C (American Institute of Physics) pp 3–14
[4] Feuillebois F, Ghalya N, Sellier A and Elasmi L 2011 *Applications of Mathematics in Technical and Natural Sciences: 3rd International Conference – AMiTaNS'11* vol 1404 ed Todorov M D and Christov C (AIP Conf. Proc.) pp 340–351
[5] Navier C L M H 1823 *Mémoires de l'Acad. des Sciences de l'Institut de France* **6** 389–416
[6] Lauga E, Brenner M P and Stone H A 2007 *Handbook of Experimental Fluid Dynamics* ed Tropea C, Foss J and Yarin A (New York: Springer-Verlag) chap 19, pp 1219–1240
[7] Beavers G and Joseph D 1967 *J. Fluid Mech.* **30** 197–207
[8] Giddings J C 1978 *Sep. Sci. Technol.* **13** 241–254 erratum: *ibid.* 14: 869–870, 1979
[9] Noel R, Gooding K, Regnier F, Ball D, Orr C and Mullins M 1978 *J. Chromatogr.* **166** 373–382
[10] Martin M 1998 *Advances in Chromatography* ed Brown P R and Grushka E (Marcel Dekker, New York) pp 1–138
[11] Ng C O 2011 *Microfluid Nanofluid* **10** 47–57
[12] Happel J and Brenner H 1973 *Low Reynolds Number Hydrodynamics* (Noordhoff, Leyden)
[13] Feuillebois F 1989 *Multiphase Science and Technology, volume 4* ed Hewitt G F, Delhaye J M and Zuber N (New York: Hemisphere Publishing Co) pp 583–798
[14] Pasol L, Martin M, Ekiel-Jeżewska M, Wajnryb E, Bławzdziewicz J and Feuillebois F 2011 *Chemical Engineering Science* **66** 4078–4089
[15] Davis A, Kezirian M and Brenner H 1994 *J.Colloid Interface Sci* **1065** 129–140
[16] Luo H and Pozrikidis C 2008 *J. Eng. Math.* **62** 1–21
[17] Dean W R and O'Neill M E 1963 *Mathematika* **10** 13–24
[18] O'Neill M E 1964 *Mathematika* **11** 67–74
[19] Loussaief H 2008 *Ecoulement de suspensions avec condition de glissement sur la paroi* Ph.D. thesis Université Pierre et Marie Curie, Paris 6 & École Polytechnique de Tunisie
[20] O'Neill M E and Bhatt B S 1991 *Q. J. Mech. appl. Math.* **44** 91–104
[21] Chaoui M and Feuillebois F 2003 *Quart. J. Mech. Applied Math.* **56** 381–410
[22] Pasol L, Sellier A and Feuillebois F 2006 *Quart. J. Mech. Applied Math.* **59** 587–614
[23] Bretherton F P 1962 *J. Fluid Mech.* **14** 284–304
[24] Sellier A, Ghalya N and Feuillebois F 2012 *J. Phys.: Conf. Ser.* **392** 012013
[25] Feuillebois F, Bazant M Z and Vinogradova O I 2009 *Phys. Rev. Lett.* **102** 026001
[26] Feuillebois F, Bazant M Z and Vinogradova O I 2010 *Phys. Rev. Lett.* **104** 159902
[27] Feuillebois F, Bazant M Z and Vinogradova O I 2010 *Phys. Rev. E* **82** 055301(R)

Journal of Physics: Conference Series **392** (2012) 012013 doi:10.1088/1742-6596/392/1/012013

Migration of a solid and arbitrarily-shaped particle near a plane slipping wall

Néjiba Ghalya[1], Antoine Sellier[2] and François Feuillebois[3]

[1] Laboratoire d'Ingénierie Mathématique, École Polytechnique de Tunisie, BP 743, 2078 La Marsa, Tunisia

[2] LadHyX. École Polytechnique, 91128. Palaiseau cédex, France

[3] LIMSI-CNRS, UPR 3251, B.P. 133, 91403 Orsay cédex, France

E-mail: `sellier@ladhyx.polytechnique.fr`

Abstract. This work is concerned with the rigid-body migration of a solid and arbitrary-shaped particle immersed in a Newtonian liquid in vicinity of a plane, motionless and impermeable wall where a Navier slip condition [1] holds. The net hydrodynamic force and torque exerted on the moving particle are obtained by appealing to a new boundary elements approach which makes use of a specific Green tensor recently determined elsewhere [2]. The advocated technique results in the treatment of a Fredholm boundary-integral equation of the first kind on the particle surface and, by contrast to earlier works in this field, makes it possible to cope with non-spherical particles. The proposed numerical implementation is benchmarked against results obtained for a sphere by using the bipolar coordinates. Preliminary new results for the friction coefficients of an non-spheroidal ellipsoid are also reported and compared with those for a volume-equivalent sphere. The variations of the friction coefficients with the slip length are analogous for both particles.

1. Introduction

The effective viscosity of a suspension bounded by a solid wall not only depends upon the shape and volume fraction of the immersed solid particles but also upon the shape and nature of the bounding wall. For sufficiently dilute suspensions it is possible to restrict, at the leading order, the problem to a single particle interacting with the wall. For applications, three different types of *motionless walls* are encountered:

(i) The quite usual solid and no-slipping wall at which one requires the fluid velocity to vanish.

(ii) The free surface at which one requires zero normal velocity and tangential stress components.

(iii) The case of a slipping and impermeable wall for which one usually employs the so-called Navier [1] boundary conditions of zero normal velocity and another relation between the non-zero tangential velocity and stress (see (3)).

Although the particle may either have a slipping or no-slipping surface, we restrict our attention to a no-slipping particle and direct the reader to [3] for the interesting case of a slipping *sphere* interacting with a slipping plane wall. In addition, we assume the boundary to be a plane wall (at least at the particle micro-scale). However, one should note that non-flat boundaries might be handled as well (as already done in Case (i)) but this challenging task would result in tremendous efforts too much involved to be treated here. In contrast to Cases (i)-(ii), Case

(iii) received a little attention in the literature (see, for instance, [4, 5, 6]) but recently attracted again attention because of experimental evidence [7]. Consequently, [8, 9] recently treated the migration of a *spherical* and solid particle in the vicinity of a plane, impermeable and slipping wall (Case (iii)) by employing the bipolar coordinates approach, as earlier done in [5, 6]. In these works [8, 9] the sphere experiences a prescribed translation or rotation *parallel* with the slipping wall or is freely-suspended in a prescribed ambient *linear* shear flow tangent to the wall. In essence the bipolar coordinates method is solely able to deal with a spherical particle and, to the authors' very best knowledge, there is unfortunately no available work coping with the challenging case of arbitrarily-shaped particles. This work therefore introduces a new approach valid for a non-spherical particle. It also reports preliminary computations for the friction coefficients of a non-spheroidal ellipsoid which are seen to differ from the ones of a volume-equivalent sphere.

2. Governing problem

We consider, as depicted in Fig. 1, a *solid* and arbitrarily-shaped particle $\mathcal{P}$, with attached point O' and smooth surface S, suspended in a Newtonian liquid with uniform density ρ_f and viscosity $\mu > 0$ occupying the $x_3 > 0$ half-space domain. We adopt Cartesian coordinates (O, x_1, x_2, x_3)

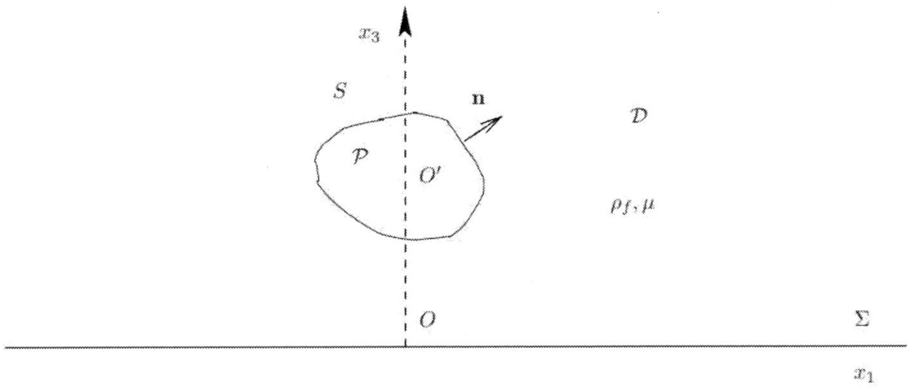

Figure 1. A *solid* and *arbitrarily-shaped* particle $\mathcal{P}$ experiencing a rigid-body motion above the $x_3 = 0$ motionless, impermeable and slipping plane wall Σ.

with origin O located on the bounding $x_3 = 0$ plane wall Σ and assume a particle rigid-body motion with prescribed translational velocity $\mathbf{U}$ (here the velocity of the attached point O') and angular velocity $\mathbf{\Omega}$. By viscosity, the particle migration induces a liquid flow with velocity $\mathbf{u}$ and pressure p in the liquid domain $\mathcal{D}$. Denoting by a and V the particle typical length-scale and the reference flow velocity we define the flow Reynolds number Re as $\mathrm{Re} = \rho_f V a/\mu$. Assuming a no-slip particle surface S, negligible inertial effects (i. e. $\mathrm{Re} \ll 1$) and a quiescent liquid far from the particle, the flow $(\mathbf{u}, p)$ is governed by the following Stokes equations, far-field behaviour and boundary condition

$$\mu\nabla^2\mathbf{u} = \nabla p \text{ and } \nabla.\mathbf{u} = 0 \text{ in } \mathcal{D}, \ (\mathbf{u}, p) \to (\mathbf{0}, 0) \text{ as } |\mathbf{OM}| \to \infty, \quad (1)$$

$$\mathbf{u} = \mathbf{U} + \mathbf{\Omega} \wedge \mathbf{O'M} \text{ on } S. \quad (2)$$

Of course, one has to supplement (1)-(2) with relevant boundary conditions on the plane wall Σ. Here, Σ is assumed to be a motionless, impermeable and *slipping surface* characterized by its prescribed slip length $\lambda > 0$ introduced in [1]. Therefore, we impose on Σ the famous and so-called Navier slip conditions

$$\mathbf{u}.\mathbf{e}_3 = 0, \ \mathbf{u}.\mathbf{e}_1 = \lambda\frac{\partial \mathbf{u}.\mathbf{e}_1}{\partial x_3} \text{ and } \mathbf{u}.\mathbf{e}_2 = \lambda\frac{\partial \mathbf{u}.\mathbf{e}_2}{\partial x_3} \text{ at } x_3 = 0. \quad (3)$$

Calculating the flow field $(\mathbf{u}, p)$ as solution of (1)-(3) and its stress tensor $\boldsymbol{\sigma}$ in the entire liquid domain for arbitrary solid particle (shape and location), rigid-body $(\mathbf{U}, \boldsymbol{\Omega})$ migration and wall slip length $\lambda > 0$ is a very challenging issue of great importance for applications. Of course, once this is achieved it is subsequently possible to calculate the net force $\mathbf{F}$ and torque $\boldsymbol{\Gamma}$ (about O') applied on the particle. Indeed, if the smooth particle surface S has unit normal $\mathbf{n}$ pointing into the liquid (as shown in Fig. 1) these quantities read

$$\mathbf{F}_h = \int_S \boldsymbol{\sigma}.\mathbf{n} dS, \ \ \boldsymbol{\Gamma}_h = \int_S \mathbf{O'M} \wedge \boldsymbol{\sigma}.\mathbf{n} dS. \tag{4}$$

3. Relevant boundary formulation and boundary-integral equations

In this section we present a boundary elements approach that provides a solution of the problem (1)-(3) for arbitrarily-shaped particles.

3.1. Usual Green tensor and boundary formulation

Let us introduce, as illustrated in Fig. 2, a so-called pole $\mathbf{y}$ in the $y_3 > 0$ half-space. We then

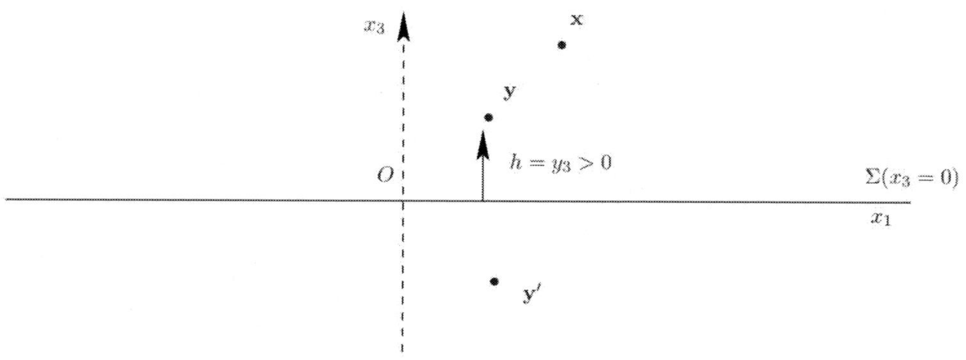

Figure 2. A pole $\mathbf{y}$ in the $y_3 > 0$ half-space and the associated image $\mathbf{y}'$ with respect to the $x_3 = 0$ plane slipping wall Σ. Here $\mathbf{x}$ is the so-called observation point.

consider for $k = 1, 2, 3$ Stokes flows with pressure $p^{(k)}$, velocity $\mathbf{v}^{(k)}$ due to a concentrated point force of strength $\mathbf{e}_k$ placed at the selected pole $\mathbf{y}$, i. e. obeying

$$\mu \nabla^2 \mathbf{v}^{(k)} = \nabla p^{(k)} - \delta_{3d}(\mathbf{x} - \mathbf{y})\mathbf{e}_k, \ \nabla.\mathbf{v}^{(k)} = 0 \text{ in the } x_3 > 0 \text{ half-space} \tag{5}$$

$$(\mathbf{u}^{(k)}, p^{(k)}) \to (\mathbf{0}, 0) \text{ as } |\mathbf{OM}| \to \infty, \tag{6}$$

with $\delta_{3d}(\mathbf{x} - \mathbf{y}) = \delta_d(x_1 - y_1)\delta_d(x_2 - y_2)\delta_d(x_3 - y_3)$ and δ_d the Dirac pseudo-function. Those flows permit one to define a so-called second-rank Green tensor $\mathbf{G}$, with Cartesian components $G_{jk}(\mathbf{x}, \mathbf{y}) = \mathbf{v}^{(k)}(\mathbf{x}, \mathbf{y}).\mathbf{e}_j$. One should note that, because there is no prescribed boundary conditions on the $x_3 = 0$ plane, the flows $(\mathbf{v}^{(k)}, p^{(k)})$ and the related Green tensor $\mathbf{G}$ are not unique. However, at that stage the basic point is that the velocity $\mathbf{u}$ solution to (1) admits in the entire liquid domain $\mathcal{D}$ and for *any* selected Green tensor $\mathbf{G}$ the following basic integral representation [10]

$$\mathbf{u}(\mathbf{x}).\mathbf{e}_j = -\int_{S \cup \Sigma} \Big\{ [\mathbf{e}_i.\boldsymbol{\sigma}.\mathbf{n}](\mathbf{y}) G_{ij}(\mathbf{y}, \mathbf{x}) - [\mathbf{u}(\mathbf{y}).\mathbf{e}_i] T_{ijk}(\mathbf{y}, \mathbf{x}) [\mathbf{n}(\mathbf{y}).\mathbf{e}_k] \Big\} dS(\mathbf{y}), \ \ \mathbf{x} \text{ in } \mathcal{D} \tag{7}$$

where we set $\mathbf{n} = \mathbf{e}_3$ on the surface Σ and the third-rank tensor $\mathbf{T} = T_{ijk}(\mathbf{y}, \mathbf{x})\mathbf{e}_i \otimes \mathbf{e}_j \otimes \mathbf{e}_k$ is the stress tensor associated with $\mathbf{G}$.

3.2. Specific Green tensor and resulting boundary formulation

As it stands, (7) shows that it is sufficient to calculate on the entire fluid boundary $S \cup \Sigma$ the velocity $\mathbf{u}$ and the surface force $\mathbf{f} = \boldsymbol{\sigma}.\mathbf{n}$. This is a nice property but unfortunately the plane wall Σ is unbounded. To circumvent this main drawback, the trick consists in adequately selecting a *specific* Green tensor $\mathbf{G}^c$ so that the integrals over the boundary Σ vanish in (7). As the reader may easily check by applying the usual reciprocal identity [11] this is achieved by selecting a Green tensor $\mathbf{G}^c$ obeying the Navier slip condition (3). In other words, the retained Green tensor $\mathbf{G}^c$ is built by using Stokes flows $(\mathbf{v}^{(k)}, p^{(k)})$ governed by (5)-(6) and the Navier slip boundary conditions

$$\mathbf{v}^{(k)}.\mathbf{e}_3 = 0, \ \ \mathbf{v}^{(k)}.\mathbf{e}_1 = \lambda \frac{\partial \mathbf{v}^{(k)}.\mathbf{e}_1}{\partial x_3} \ \text{ and } \ \mathbf{v}^{(k)}.\mathbf{e}_2 = \lambda \frac{\partial \mathbf{v}^{(k)}.\mathbf{e}_2}{\partial x_3} \quad \text{on } \Sigma. \tag{8}$$

Using (7) for the Green tensor $\mathbf{G}^c$ and its associated stress tensor $\mathbf{T}^c$ then provides the more pleasant velocity representation

$$\mathbf{u}(\mathbf{x}).\mathbf{e}_j = -\int_S \Big\{ [\mathbf{e}_i.\boldsymbol{\sigma}.\mathbf{n}](\mathbf{y}) G^c_{ij}(\mathbf{y},\mathbf{x}) - [\mathbf{u}(\mathbf{y}).\mathbf{e}_i] T^c_{ijk}(\mathbf{y},\mathbf{x})[\mathbf{n}(\mathbf{y}).\mathbf{e}_k] \Big\} dS(\mathbf{y}), \ \ \mathbf{x} \text{ in } \mathcal{D}. \tag{9}$$

Another improvement is also obtained by noting that we are confining ourselves in this work to a velocity field $\mathbf{u}$ which is, by virtue of the boundary condition (2), a rigid-body motion on the particle surface S. As shown in [12], the second integral on the right-hand side of (9) (the so-called double-layer term) then vanishes! Accordingly, one ends up for the Stokes flow $(\mathbf{u}, p)$ governed by (1)-(2) with the simple integral representation

$$\mathbf{u}(\mathbf{x}).\mathbf{e}_j = -\int_S [\mathbf{e}_i.\boldsymbol{\sigma}.\mathbf{n}](\mathbf{y}) G^c_{ij}(\mathbf{y},\mathbf{x}) dS(\mathbf{y}), \ \ \mathbf{x} \text{ in } \mathcal{D} \tag{10}$$

which provides in the entire liquid domain the velocity $\mathbf{u}$ solely in terms of the surface force $\mathbf{f} = \boldsymbol{\sigma}.\mathbf{n}$ exerted by the Stokes flow $(\mathbf{u}, p)$ on the particle surface. In practice, the required vector $\mathbf{f}$ is gained by letting the point $\mathbf{x}$ in (10) tend onto the surface S. This procedure is valid because, as seen in §3.3, each Cartesian component $G^c_{ij}(\mathbf{y},\mathbf{x})$ is weakly-singular as $\mathbf{x}$ approaches $\mathbf{y}$. As a result, one gets the following Fredholm boundary-integral equation of the first kind

$$[\mathbf{U} + \boldsymbol{\Omega} \wedge \mathbf{O'M}].\mathbf{e}_j = -\int_S [\mathbf{f}.\mathbf{e}_i](\mathbf{y}) G^c_{ij}(\mathbf{y},\mathbf{x}) dS(\mathbf{y}) \text{for } \mathbf{x} = \mathbf{OM} \text{ on } S. \tag{11}$$

In summary, the advocated treatment of the Stokes problem (1)-(3) appeals to the following steps:

(i) Obtain the previously-introduced specific Green tensor $\mathbf{G}^c$.

(ii) Invert the boundary-integral equation (11) to get the surface force $\mathbf{f}$ exerted by the flow $(\mathbf{u}, p)$ on the particle surface.

(iii) Evaluate the net force and torque $\mathbf{F}$ and $\boldsymbol{\Gamma}$ given by (4) from the knowledge of $\mathbf{f}$.

(iv) Compute, if needed, the velocity field $\mathbf{u}$ in the liquid domain by employing the nice integral representation (10).

3.3. Determination of the specific Green tensor

The key step for our procedure is the determination of the specific Green tensor $\mathbf{G}^c$. More precisely, we look at the Cartesian component $G^c_{jk}(\mathbf{x},\mathbf{y}) = \mathbf{v}^{(k)}(\mathbf{x},\mathbf{y}).\mathbf{e}_j$ for pole $\mathbf{y}$ and a so-called observation point $\mathbf{x}$ located in the half-space domain above the plane wall Σ (see Fig. 2) and Stokes flows $(\mathbf{v}^{(k)}, p^{(k)})$ fulfilling (5)-(6) and (8). Here we briefly present the results and direct for further details the reader to [2]. First, we introduce the well-known [12] free-space Oseen-Burgers Green tensor tensor $\mathbf{G}^{\text{free-space}}$ having Cartesian components and associated pressure $p^{free-space,(k)}$

$$G_{jk}^{\text{free-space}}(\mathbf{x},\mathbf{y}) = \frac{1}{8\pi\mu}\left\{\frac{\delta_{jk}}{|\mathbf{x}-\mathbf{y}|} + \frac{[(\mathbf{x}-\mathbf{y}).\mathbf{e}_j][(\mathbf{x}-\mathbf{y}).\mathbf{e}_k]}{|\mathbf{x}-\mathbf{y}|^3}\right\}, \tag{12}$$

$$p^{free-space,(k)}(\mathbf{x},\mathbf{y}) = \frac{(\mathbf{x}-\mathbf{y}).\mathbf{e}_k}{4\pi} \tag{13}$$

with δ the Kronecker delta symbol. Then, we use the decompositions

$$G_{jk}^{c}(\mathbf{x},\mathbf{y}) = G_{jk}^{\infty}(\mathbf{x},\mathbf{y}) - G_{jk}^{\infty}(\mathbf{x},\mathbf{y}') + w_j^{(k)}(\mathbf{x},\mathbf{y}), \tag{14}$$

$$p^{c,(k)}(\mathbf{x},\mathbf{y}) = p^{\infty,(k)}(\mathbf{x},\mathbf{y}) - p^{\infty,(k)}(\mathbf{x},\mathbf{y}') + s^{(k)}(\mathbf{x},\mathbf{y}) \tag{15}$$

with, as depicted in Fig.2, $\mathbf{y}'$ the symmetric of the pole $\mathbf{y}$ with respect to the plane slipping wall Σ. The Fourier transforms of functions $w_j^{(k)}$ and $s^{(k)}$ are obtained as detailed in [2]. Applying the inverse Fourier transform finally provides $w_j^{(k)}$ and $s^{(k)}$. Setting $h = \mathbf{y}.\mathbf{e}_3, x_3 = \mathbf{x}.\mathbf{e}_3, R_3' = x_3 + h, R_j = (\mathbf{x}-\mathbf{y}).\mathbf{e}_j$ for $j = 1,2$ and $\rho = \{R_1^2 + R_2^2\}^{1/2}$ one gets (see [2])

$$\frac{4\pi\mu}{h} w_1^{(3)} = \frac{R_1}{\rho}\left\{\int_0^\infty \frac{\xi e^{-R_3'\xi} J_1(\rho\xi) d\xi}{1+2\lambda\xi} - x_3 \int_0^\infty \frac{\xi^2 e^{-R_3'\xi} J_1(\rho\xi) d\xi}{1+2\lambda\xi}\right\}, \tag{16}$$

$$\frac{4\pi\mu}{h} w_2^{(3)} = \frac{R_2}{\rho}\left\{\int_0^\infty \frac{\xi e^{-R_3'\xi} J_1(\rho\xi) d\xi}{1+2\lambda\xi} - x_3 \int_0^\infty \frac{\xi^2 e^{-R_3'\xi} J_1(\rho\xi) d\xi}{1+2\lambda\xi}\right\}, \tag{17}$$

$$\frac{4\pi\mu}{h} w_3^{(3)} = -x_3 \int_0^\infty \frac{\xi^2 e^{-R_3'\xi} J_0(\rho\xi) d\xi}{1+2\lambda\xi}, \tag{18}$$

$$\begin{aligned}\frac{4\pi\mu}{h} w_1^{(1)} &= [2\lambda + x_3(1+2\lambda/h)]\left\{(\frac{R_1}{\rho})^2 \int_0^\infty \frac{\xi^2 e^{-R_3'\xi} J_2(\rho\xi) d\xi}{1+2\lambda\xi} - \frac{1}{\rho}\int_0^\infty \frac{\xi e^{-R_3'\xi} J_1(\rho\xi) d\xi}{1+2\lambda\xi}\right\} \\ &+ \frac{2\lambda^2}{h}[\frac{1}{\rho}\int_0^\infty \frac{\xi e^{-R_3'\xi} J_1(\rho\xi) d\xi}{(1+2\lambda\xi)(1+\lambda\xi)} - (\frac{R_2}{\rho})^2 \int_0^\infty \frac{\xi^2 e^{-R_3'\xi} J_2(\rho\xi) d\xi}{(1+2\lambda\xi)(1+\lambda\xi)}] \\ &+ \frac{2\lambda}{h}\int_0^\infty \frac{\xi e^{-R_3'\xi} J_0(\rho\xi) d\xi}{1+2\lambda\xi},\end{aligned} \tag{19}$$

$$\begin{aligned}\frac{4\pi\mu}{h} w_2^{(1)} &= \frac{R_1 R_2}{\rho^2}\left\{[x_3 + 2\lambda(1+x_3/h)]\int_0^\infty \frac{\xi^2 e^{-R_3'\xi} J_2(\rho\xi) d\xi}{1+2\lambda\xi}\right. \\ &\left. + \frac{2\lambda^2}{h}\int_0^\infty \frac{\xi^2 e^{-R_3'\xi} J_2(\rho\xi) d\xi}{(1+2\lambda\xi)(1+\lambda\xi)}\right\}\end{aligned} \tag{20}$$

$$\frac{4\pi\mu}{h} w_3^{(1)} = \frac{R_1}{\rho}\left\{x_3(1+2\lambda/h)\int_0^\infty \frac{\xi^2 e^{-R_3'\xi} J_1(\rho\xi) d\xi}{1+2\lambda\xi} + \frac{\rho}{(\rho^2 + R_3'^2)^{3/2}}\right\} \tag{21}$$

with J_m the Bessel function of positive integer order m. The quantities $w_k^{(2)}$ are obtained by switching superscripts or subscripts 1 and 2 in (19)-(21).

4. Numerical implementation and preliminary numerical results

This section briefly presents the numerical method. It also provides benchmark tests against the results obtained elsewhere [5, 6, 8, 9] for a spherical particle and reports preliminary numerical results for a non-spherical particle.

4.1. Numerical method and benchmark tests

A collocation method [13, 14] is implemented to numerically invert the boundary-integral equation (11). For a sake of accuracy, we actually use on the particle surface a $N-node$ mesh consisting of $6-node$ curved triangular boundary elements. The discretized counter-part of (11) is a linear system $AX=Y$ with a $3N\times 3N$ dense and non-symmetric influence square matrix A with coefficients calculated as explained in [14]. The reader should note that each integral arising on the right-hand sides of identities (16)-(21) must be accurately computed. For a sake of conciseness the employed method is however not detailed here and the reader is directed to [15] for further details. Finally, the system $AX=Y$ is solved by Gaussian elimination.

Comparisons are made against the results previously-obtained in the literature, using the quite different method of bipolar coordinates [5, 6, 8], for a spherical particle with radius a and center O' such that $\mathbf{OO'}=l\mathbf{e}_3$ and $l>a$. By linearity and also for symmetry reasons, the net force $\mathbf{F}$ and net torque $\mathbf{\Gamma}$ (about the sphere center O') exerted by the Stokes flow about the moving sphere with a rigid-body motion $(\mathbf{U},\mathbf{\Omega})$ solely depend upon six so-called friction coefficients $f_{33}, c_{33}, f_{11}, c_{12}, f_{21}$ and c_{22} defined as follows

$$\mathbf{F}=-6\pi\mu a f_{33}\mathbf{U} \quad \text{and} \quad \mathbf{\Gamma}=\mathbf{0} \quad \text{for} \quad \mathbf{U}\wedge\mathbf{e}_3=\mathbf{\Omega}=\mathbf{0}, \tag{22}$$

$$\mathbf{\Gamma}=-8\pi\mu a^3 c_{33}\mathbf{\Omega} \quad \text{and} \quad \mathbf{F}=\mathbf{0} \quad \text{for} \quad \mathbf{\Omega}\wedge\mathbf{e}_3=\mathbf{U}=\mathbf{0}, \tag{23}$$

$$\mathbf{F}=-6\pi\mu a f_{11}\mathbf{U} \quad \text{and} \quad \mathbf{\Gamma}=-8\pi\mu a^2 c_{12}\mathbf{U}\wedge\mathbf{e}_3 \quad \text{for} \quad \mathbf{U}\wedge\mathbf{e}_1=\mathbf{\Omega}=\mathbf{0}, \tag{24}$$

$$\mathbf{\Gamma}=-8\pi\mu a^3 c_{22}\mathbf{\Omega} \quad \text{and} \quad \mathbf{F}=-6\pi\mu a^2 f_{21}\mathbf{\Omega}\wedge\mathbf{e}_3 \quad \text{for} \quad \mathbf{\Omega}\wedge\mathbf{e}_2=\mathbf{U}=\mathbf{0}. \tag{25}$$

One should note that using the reciprocal identity [10, 11] and the above definitions gives the relation $f_{21}=4c_{12}/3$ so that one ends up with five coefficients.

Computed values of f_{11}, c_{12}, f_{21} and c_{22} are compared with the predictions of [6, 8] for different $N-node$ meshes on the sphere surface in Tab. 1. Clearly, there is a nice agreement between the bipolar coordinates method and the advocated and quite different boundary elements approach.

Table 1. Computed friction coefficients f_{11}, c_{12}, f_{21} and c_{22} for $l/a=1.5$ and $\lambda=1$ versus the number N of nodal points put on the sphere surface.

N	f_{11}	c_{12}	f_{21}	c_{22}
74	1.28628	-0.026335	-0.034849	1.03624
242	1.28981	-0.026200	-0.035023	1.03560
1058	1.29020	-0.026285	-0.035054	1.03548
[6, 8]	1.29023	-0.026293	-0.035057	1.03548

In a similar fashion, comparisons with [5] for the friction coefficients f_{33} and c_{33} are given in Tab. 2. One should note that values given in [5] have been actually corrected by coding the equations obtained in [5] using a 16-digit accuracy. Again, a nice convergence towards the (corrected) values obtained by the bipolar coodinates procedure is observed.

Table 2. Computed friction coefficients f_{33} and c_{33} for $l/a=1.543$ and $\lambda=2.5$ versus the number N of nodal point spread on the sphere surface.

	$N=74$	$N=242$	$N=1058$	[4]	[4]Corrected
f_{33}	2.11545	2.12474	2.12572	2.11843	2.12568
c_{33}	0.99321	0.98947	0.98934	0.98835	0.98935

4.2. Numerical results for a solid ellipsoidal particle

By contrast to the bipolar coordinates approach restricted to a spherical particle, the advocated boundary method is able to cope with non-spherical particles. It thus permits us to investigate to which extent the force and torque exerted on a solid particle moving near the slipping wall depend upon the particle shape. This is achieved in the present work by comparing the results for the sphere with radius a and for a volume-equivalent ellipsoidal particle with center of volume O' and semi-axis a_i parallel with $\mathbf{e}_i$ for $i = 1, 2, 3$. More precisely, we take $a_1 = 1.2a, a_2 = 1/a_1, a_3 = a$ and set $\mathbf{OO'} = l\mathbf{e}_3$ with $l > a_3 = a$. Accordingly, for given l/a the particle-wall gap is $l-a$ both for the sphere and the ellipsoid. For the selected ellipsoid orientation the relations (22)-(25) still hold with new friction coefficients $f_{33}, c_{33}, f_{11}, c_{12}, f_{21}$ and c_{22} for the ellipsoid. One should note that additional friction coefficients, obtained for $\mathbf{U} \wedge \mathbf{e}_2 = \mathbf{\Omega} = \mathbf{0}$ and for $\mathbf{\Omega} \wedge \mathbf{e}_1 = \mathbf{U} = \mathbf{0}$, are found for the ellipsoid. Here we however confine our comparisons to the coefficients $f_{33}, c_{33}, f_{11}, c_{22}$ and $f_{21}(= 4c_{12}/3)$ which are computed, versus l/a and for different slip lengths λ, both for the sphere and the ellipsoid using $N = 1058$ collocation points on each particle surface.

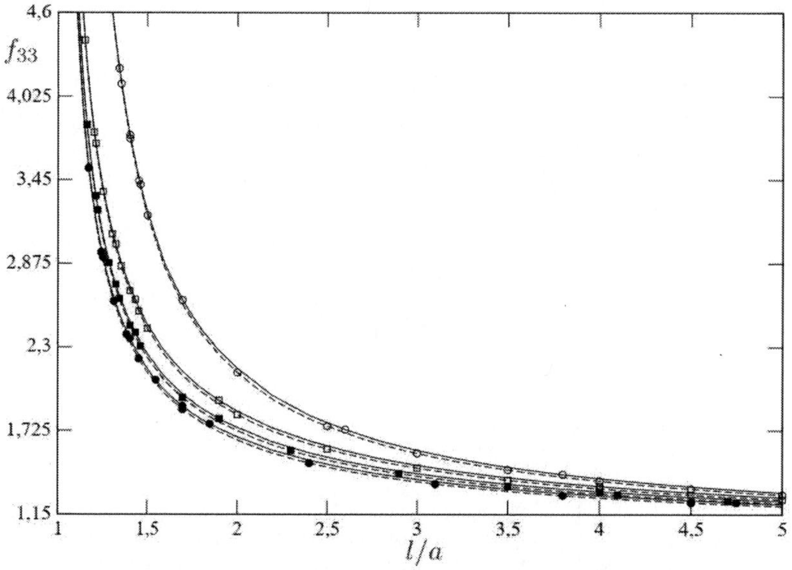

Figure 3. Compared friction coefficients f_{33} for a sphere (dashed lines) and the ellipsoid (solid lines) for different slip lengths: from top to bottom ($\circ$) for $\lambda/a = 0$, ($\square$) for $\lambda/a = 0.7$, ($\blacksquare$) for $\lambda/a = 2$, ($\bullet$) for $\lambda/a = 5$.

Computed coefficient f_{33} (obtained when the particle translates normal to the slipping wall) is plotted, both for the ellipsoid and the volume-equivalent sphere, in Fig. 3 versus the separation parameter l/a for different normalized slip lengths λ/a. It is seen that f_{33} is nearly the same for the two particles: it increases as the particle approaches the slipping wall and also decreases as the normalized slip length λ/a increases. One should also note that $f_{33} > 1$ whatever the particle location and the ratio λ/a (this is also the case for a free surface obtained by letting λ/a tend to infinity).

The coefficient f_{11} is shown in Fig. 4 again versus the separation parameter l/a and for different normalized slip lengths λ/a. For the ellipsoid this friction coefficient is always weaker than for the equivalent sphere. For both particles f_{11} decreases as λ/a increases for a prescribed location l/a. In contrast to the previously-discussed case of f_{11}, note that for λ/a sufficiently large the function f_{11} does not monotically decrease as l/a increases (see the curves associated with the value $\lambda/a = 5$). This means, not surprisingly, that a larger slip reduces the friction for a particle translating parallel to a close slipping wall.

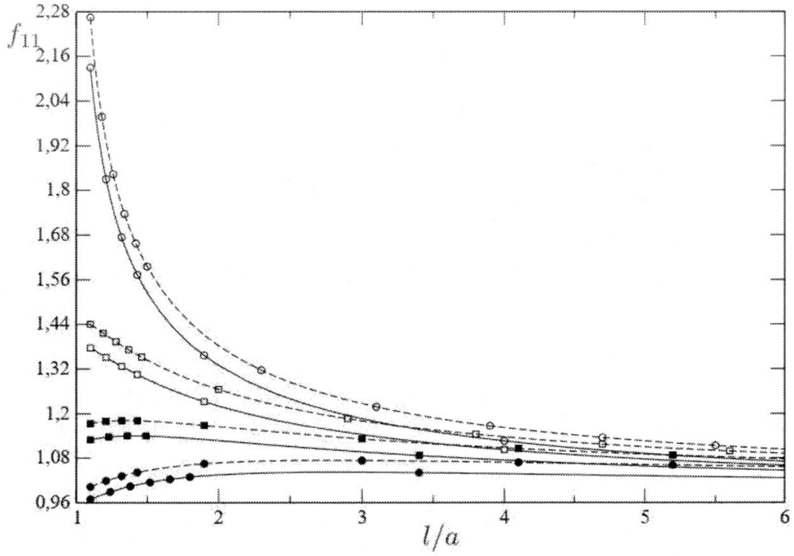

Figure 4. Compared friction coefficients f_{11} for a sphere (dashed lines) and the ellipsoid (solid lines) for different slip lengths: from top to bottom (∘) for $\lambda/a = 0$, (□) for $\lambda/a = 0.7$, (■) for $\lambda/a = 2$, (•) for $\lambda/a = 5$.

Fig 5. displays the friction coefficients c_{22} versus l/a for different values of λ/a. This coefficient is larger than unity, behaves as f_{33} for each particle and is clearly shape-dependent. The coefficient c_{33} addressed in Fig. 6 is also larger for the ellipsoid. When λ/a is large enough c_{33} decreases with l/a and might be even less than unity for a particle sufficiently close to the wall (this arises in Fig. 6 for the sphere but would also be observed for the ellipsoid for λ/a exceeding a critical value larger than 5).

As seen in Fig. 7, the coefficient c_{12} is of weak magnitude and, depending upon the value of λ/a, either increases or decreases as the particle approaches the slipping wall. For a given particle location note that c_{12} vanishes and becomes negative for a critical ratio λ/a which strongly depends on the particle shape.

5. Conclusions

A new boundary elements approach consisting of the treatment of one Fredholm boundary-integral equation of the first kind on the particle surface has been proposed, implemented and benchmarked against results available for a spherical particle. The advocated procedure appeals to a specific Green tensor obtained in [2] and holds for arbitrary-shaped particles. Preliminary comparisons for a sphere and a volume-equivalent non-spheroidal ellipsoid reveal that all friction coefficients show the same trends versus the slip length for both particles while some of the

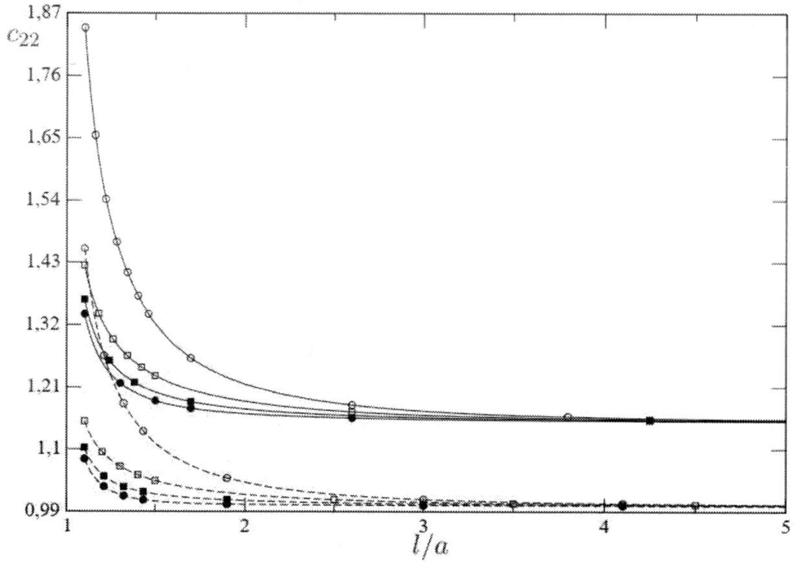

Figure 5. Compared friction coefficients c_{22} for a sphere (dashed lines) and the ellipsoid (solid lines) for different slip lengths: from top to bottom (∘) for $\lambda/a = 0$, (□) for $\lambda/a = 0.7$, (■) for $\lambda/a = 2$, (•) for $\lambda/a = 5$.

coefficients are deeply sensitive to the particle shape.

[1] Navier C. L. M. H. 1823 *Mémoires de l'Académie Royale des Sciences de l'Institut de France*, **VI**, 389-416.
[2] Sellier A. and Gahlia N. Green tensor for a general non-isotropic slip condition 2011 *CMES*,**78(1)**, 25-50.
[3] Luo H. and Pozrikidis C. 2008 *J. Eng. Math.*,**62(1)**, 1-21.
[4] Churaev N.V., Sobolev V.D. and Somov A. N. 1984 *J. Colloid Int. Sci.*,**97**, 574-581.
[5] O'Neill M. E. and Bhatt B. S. 1991 *Q. J. Mech. Appl. Math.*,**44**, 91-104.
[6] Davis A.M.J., Kezirian M.T. and Brenner H. 1994 *J.Colloid Interface Sci.*,**1065**, 129-140.
[7] Baudry J., Charlaix E., Tonck A. and Mazuyer D. 2001 *Langmuir*,**17**, 5232-5236.
[8] Loussaief H. 2008 Ecoulement de suspensions avec condition de glissement sur la paroi. *PHD Thesis.* Laboratory of Engineering Mathematics, Polytechnic School of Tunisia, BP 743, La Marsa, Tunisia.
[9] Feuillebois F., Loussaief H. and Pasol L. 2009 Particles in Creeping Flow Near a Slip Wall. *CP1186, Applications of Mathematics in Technical and Natural Sciences* Editors M. D. Todorov and C.I. Christov. American Institute of Physics, 3-14.
[10] Kim S. and Karrila S. J. 1991 *Microhydrodynamics: principles and selected applications* (Butterworth).
[11] Happel J. and Brenner H. 1973 *Low Reynolds number hydrodynamics*, Martinus Nijhoff.
[12] Pozrikidis C. 1992 *Boundary integral and singularity methods for linearized viscous flow*, Cambridge University Press.
[13] Bonnet M. 1999 *Boundary Integral Equation Methods for Solids and Fluids*, John Wiley & Sons Ltd.
[14] Sellier A. 2011 Boundary Element Technique for slow viscous flows about particles. *In boundary Element Methods in Engineering And Sciences. Computational and Experimental Methods In Structures - Vol 4.* Editors M.H. Aliabadi and P. H. Wen. Imperial College Press.
[15] Sellier A and Pasol, L. 2006 Sedimentation of a Solid Particle Immersed in a Fluid Film *CMES*,**16(3)**, 187-196.

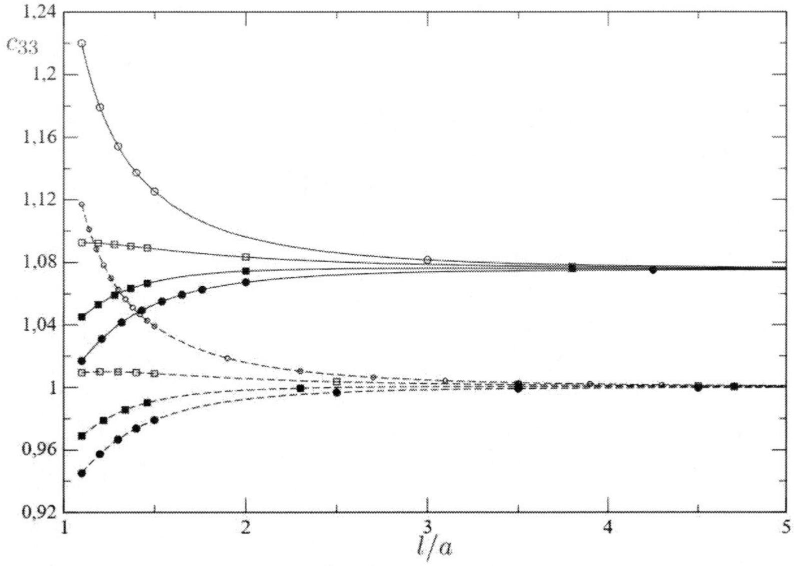

Figure 6. Compared friction coefficients c_{33} for a sphere (dashed lines) and the ellipsoid (solid lines) for different slip lengths: from top to bottom (∘) for $\lambda/a = 0$, (□) for $\lambda/a = 0.7$, (■) for $\lambda/a = 2$, (•) for $\lambda/a = 5$.

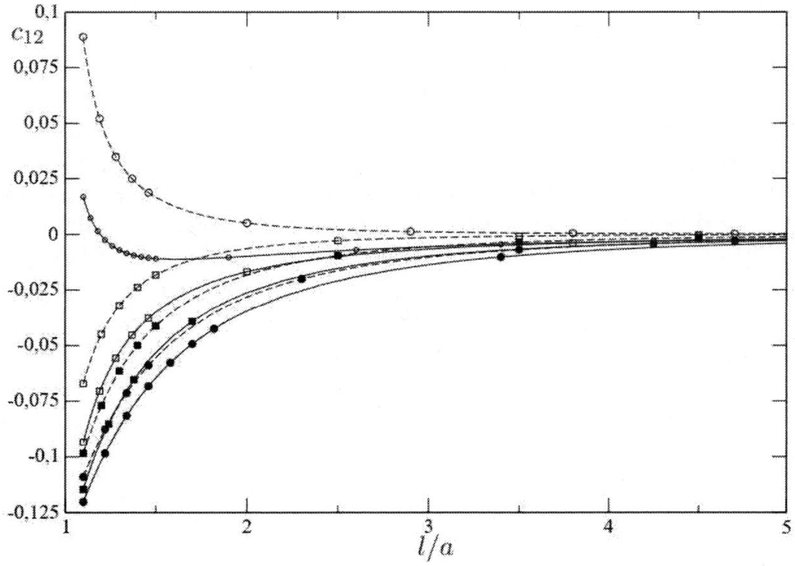

Figure 7. Compared friction coefficients c_{12} for a sphere (dashed lines) and the ellipsoid (solid lines) for different slip lengths: from top to bottom (∘) for $\lambda/a = 0$, (□) for $\lambda/a = 0.7$, (■) for $\lambda/a = 2$, (•) for $\lambda/a = 5$.

Journal of Physics: Conference Series **392** (2012) 012014 doi:10.1088/1742-6596/392/1/012014

Analysis of wall effect on the process of diffusion of nanoparticles in a microchannel

K. Zembrzycki*, **S. Błoński, T. A. Kowalewski**

Institute of Fundamental Technological Research, Polish Academy of Sciences, IPPT PAN, Pawinskiego 5B, 02-106 Warsaw, Poland

E-mail: *kzem@ippt.gov.pl

Abstract. In this preliminary work we introduce a new method for verification of the no-slip boundary condition on the liquid-solid interface, by analyzing variations in Brownian motion coefficients of colloidal nanoparticles as a function of distance from the wall. The experimental investigations are performed in a small channel using an epi-fluorescent microscope. For precise measurements close to the wall an evanescent wave illumination is used. The experimental data obtained for $300nm$ particles gave us evidence of relatively large $(0.3\mu m)$ slip length. The experiments are supplemented by two-dimensional Molecular Dynamics simulations.

1. Introduction

Recently, the classical no-slip hypothesis for the tangential velocity of liquid adjacent to the solid surface became questioned for the flow in micro- and nanofluidic devices. The topic is of fundamental interest and has potential practical consequences in many areas of applied sciences. For small-scale geometries this slip can be significant permitting for nearly frictionless flow in 2nm carbon nanotubes [1].

Hence, several experimental and numerical studies has been performed to demonstrate and explain presence of the molecular-scale slip at the fluid-solid interface (see the reviews by Neto *et al.* [2], Lauga *et al.* [3]). The key factor is that the continuum fluid dynamics still holds within the interior of the liquid and the slip is due entirely to discontinuity in speed between the wall and adjacent liquid layer. Molecular Dynamics simulations [4] gave some insight into the problem, suggesting localized hoping of adjacent to the wall liquid molecules or discontinuous motion of the whole molecular layers.

Slip is usually viewed from the averaged flow properties over all liquid molecules near the solid boundary and expressed as slip length, which is determined by extrapolating velocity profile beyond the channel wall (Fig. 1).

The amount of the slip is quantified in terms of the shear rate and local flow velocity using following formula:

$$U_s = \lambda\dot{\gamma},$$
$$\dot{\gamma} = \frac{dU(z)}{dz}, \tag{1}$$

where:
U - fluid velocity,
U_s - fluid velocity at the wall,
$\dot{\gamma}$ - shear rate,
λ - slip length.

From the theoretical point of view (Molecular Dynamics studies [4]) the slip length strongly depends on the shear rate and varies from fractions of nanometer to few nanometers only. Experimental studies indicate much larger slip variation, covering range from several nanometers to several micrometers (see collection of data by Lauga *et al.* [3]). This ambiguity is mainly due to different surface treatments and different liquid/solid interactions like wetting - non-wetting. But perhaps the main source of differences comes from experimental difficulties in performing accurate velocity measurements of the flow at molecular distances from the wall [8, 9, 10, 11, 12].

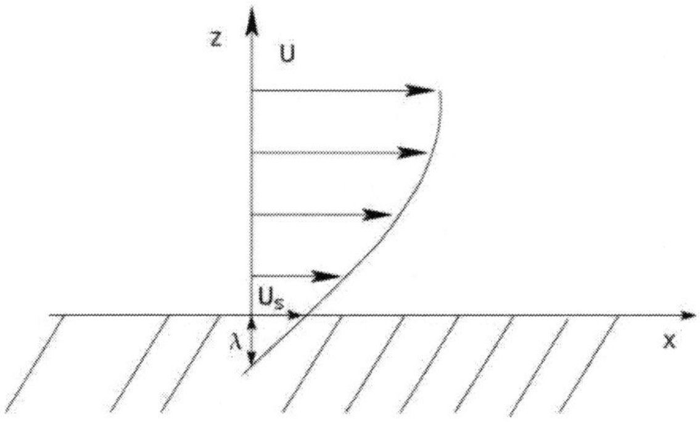

Figure 1. Navier slip law.

To avoid problems appearing in reconstructing flow velocity profile in nanometric liquid layers, we propose to evaluate how the slip/no-slip boundary conditions affect behavior of a Brownian particle performing its chaotic motion close to the wall. In physics Brownian motion has been a continuous inspiration for understanding diverse transport processes. It was found that depending on fluid properties, molecular size ratio [5], or spatial constrains [6, 7] stochastic motion of Brownian particle changes. By evaluating deviations of the diffusion coefficient one obtains a micro-scale sensor of local liquid parameters and flow conditions.

In the fluid bulk, the diffusivity D of an isolated particle follows the Stokes-Einstein relation, which balance the fluid thermal energy with the particles hydrodynamic mobility. In the vicinity of the wall hydrodynamic mobility changes and depends on the direction of the particle motion. The diffusion coefficients $(D_\perp, D_\|)$ for a single particle close to the wall can be defined by the following expressions:

$$D_\perp = \beta_\perp D, D_\| = \beta_\| D, \tag{2}$$

$$D = \frac{k_B T}{3\pi\mu d_p} \tag{3}$$

where:
$D_\perp$ - diffusion coefficient towards the wall,
$D_\parallel$ - diffusion coefficient parallel to the wall,
$\beta_\perp$ - perpendicular variation,
$\beta_\parallel$ - parallel variation,
D - bulk diffusion coefficient,
k_B - boltzman constant,
T - temperature,
μ - viscosity,
d_p - particle diameter.

The hydrodynamic mobility of the particle close to the wall depends on the boundary conditions imposed for the velocity field at the wall. The slip or no-slip boundary conditions change apparent diffusivity of the particle. Lauga and Squires [13] evaluated this effect by deriving general formula for the diffusion coefficient of a single particle as a function of the distance h from the wall, and the slip length λ assumed for the kinematic boundary conditions:

$$\beta_\perp = 1 - \frac{3d_p}{8h}\left[1 + 2I(Kn)\right] + O\left(\frac{d_p^3}{8h^3}\right), \tag{4}$$

$$\beta_\parallel = 1 + \frac{3d_p}{16h}\left[1 - 2J(Kn)\right] + O\left(\frac{d_p^3}{8h^3}\right), \tag{5}$$

where: h - is distance from particle center to the wall. Knudsen number $Kn = \lambda/h$ represents nondimensional measure of the slip length.
Functions I and J are defined by the following integrals [13]:

$$I(Kn) = \int_0^\infty \frac{x^2}{1 + 2xKn} e^{-2x} dx,$$

$$J(Kn) = \int_0^\infty \frac{(1 + xKn)(1 - x)^2 + 2(1 + 2xKn)}{(1 + xKn)(1 + 2xKn)} e^{-2x} dx.$$

Here, the above formulation is used to evaluate effect of the wall on the Brownian motion of nanoparticles suspended in water. The same procedure is used for the data obtained from the numerical simulations using Molecular Dynamics and for the experimental study. Having distance h of the analyzed particles from the wall and their diffusion coefficients (5), by inverting the I and J integrals we may evaluate the Knudsen number Kn, hence the slip length λ for parallel and perpendicular to the wall flow velocity components. In this way the slip length as a function of distance from the wall h is obtained.

Figure 2 shows variation of I and J functions with Knudsen number Kn. It can be noticed that the main effect of the slip length can be found for Kn in the range of 0.1 to 10. In this range a relatively small change of Kn results with a quite strong variation of both functions involved in the correction of the diffusivity constant. The largest effect can be expected for parallel component of the diffusion constant ($J(Kn)$ variation).

2. Simulations

The Brownian diffusion of 24 nm carbon nanoparticle suspended in water in an infinite channel formed between two quartz walls (Fig. 3) was simulated by the molecular dynamics code LAMMPS (Large-scale Atomic/Molecular Massively Parallel Simulator). The computations were performed for the square domain of $310.5nm$ size and periodic boundary

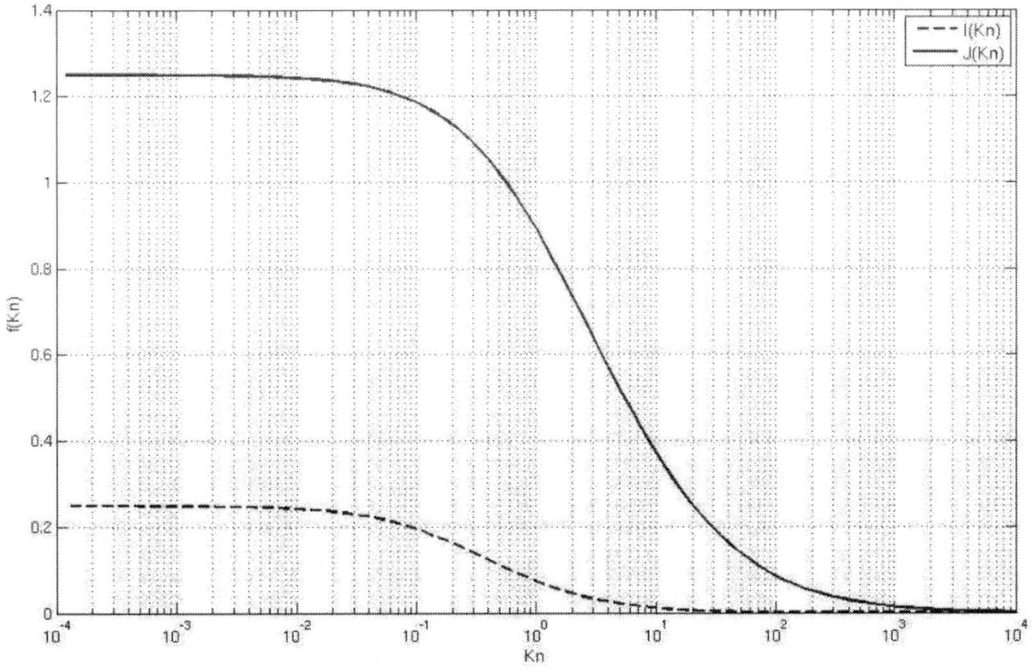

Figure 2. Variation with the Knudsen number Kn of the functions I and J involved in the correction to perpendicular and parallel diffusivity close to the wall.

conditions for the two side walls. All interactions between particles were described by the Lennard-Jones potential:

$$U(r) = 4 \in \left[\left(\frac{\sigma}{r}\right)^{12} - \left(\frac{\sigma}{r}\right)^{6} \right] \tag{6}$$

where:
$\in$ - potential well depth,
σ - distance at which the potential is zero,
r - distance between molecules.

The colloidal particles as well as the walls were defined as independent rigid bodies. This means that at each time step the total force on each rigid body is computed as the sum of the forces on its constituent particles and the coordinates and velocities of the atoms in each body are updated so that the body moves and rotates as a single entity. The following potential parameters where used for all simulations:

$$\in_{H_2O}= 0.155\frac{kcal}{mol}, \sigma_{H_2O} = 3.1536\mathring{A} - \text{liquid particles,}$$

$$\in_{C}= 0.102\frac{kcal}{mol}, \sigma_{C} = 3.35\mathring{A} - \text{colloid particles,}$$

$$\in_{SiO_2}= 0.9882\frac{kcal}{mol}, \sigma_{SiO_2} = 2.503\mathring{A} - \text{wall particles.}$$

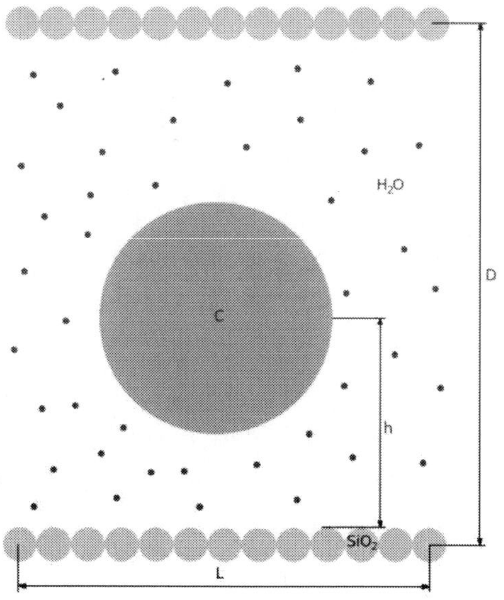

Figure 3. A schematic picture of the MD simulation domain with one $24nm$ colloidal carbon particle suspended in water between two quartz walls. $L = D = 310.5nm$.

For calculating the parameter of interactions between different types of particles (i and j) the Lorentz-Berthelot mixing rules were used:

$$\sigma_{ij} = \frac{\sigma_i + \sigma_j}{2}, \tag{7}$$

$$\in_{ij} = \sqrt{\in_i \in_j}. \tag{8}$$

The cut off distance was set to 3σ. The time step was set to $0.1fs$ and the number of time steps was $2 * 10^6$. In each simulation the number of particles was nearly 10^6. Data was gathered after the simulation reached equilibrium state, which was after about $5 * 10^5$ time steps. A standard Velocity-Verlet integrator was used to calculate forces.

To calculate the diffusion coefficient the following formula was used:

$$D = \frac{< s^2 >}{A\Delta t}, \tag{9}$$

where:
Δt - time, s - distance,
A - dimension coefficient, equals 2, 4, or 6 for $1D$, $2D$ and $3D$ problem.

Several simulations were performed to evaluate Brownian motion of colloidal particles placed close to the wall. By calculating the parallel displacement and by knowing the particle distance from the wall we could estimate the average slip length which is given by (5):

$$\beta_{\|} = 0.52,\ h = 12.3nm\ \lambda = 3.25nm$$

The outcome of this preliminary numerical analysis suggests relatively small slip length. Generally, it is in accordance with Molecular Dynamics simulations reported for the channel flow [3, 4]. However, due to the obvious constrains of our numerical model (short physical simulation time, two-dimensional domain, frozen wall molecules) the obtained value needs further verification. Additional long time simulations with particles placed at different distances from the wall are continued.

3. Experiment

In the experimental study we analyzed the Brownian motion of fluorescent particles with a diameter of $300nm$ suspended in water kept between two quartz walls of a microchannel. The distance between the lower and upper wall of the channel was $500\mu m$. To obtain precise measurements very close to the upper wall evanescent wave was generated by total internal reflection of the illuminating laser beam (Fig. 4.). Light undergoes total internal reflection (TIR) at a refractive index interface between a dense medium with high refractive index (glass) and a less dense medium (water) when the angle of incidence α exceeds the critical value. Then part of the light penetrates into the less dense medium and propagates parallel to the interface with an intensity that decays exponentially with the normal distance from the interface. This evanescent wave illumination has been used extensively in the life sciences. Recently it was rediscovered in microfluidics for near wall flow measurements [3]. The main advantage of the method is possibility to reduce the depth of focus of the system below 100nm. Hence, we use TIR method to obtain images only of particles which are in the vicinity of the wall. All other particles suspended in the liquid stay out of the illuminating sheet and are practically invisible for the acquisition system.

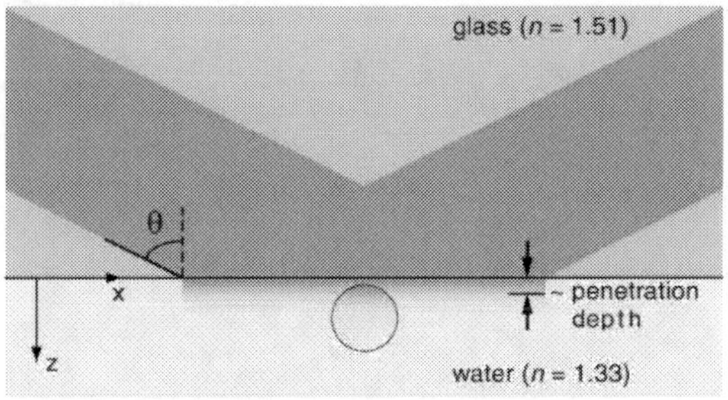

Figure 4. Graphical representation of evanescent wave illumination.

The cavity formed between two quartz plates once filled with water suspension of fluorescent particles was sealed and mounted on the microscope table. A quartz prism was attached to the upper wall to couple the laser light beam into the channel. The penetration depth is a function of incidence beam angle on the prism as it is illustrated on Figure 5. A precise mechanical system equipped with movable mirrors was used to set appropriate incidence angle. Having fixed light incidence angle the penetration depth in the setup could be calculated using the following equations:

$$d = \frac{\Lambda}{4\pi}\left[\eta_1^2 sin^2\left(\frac{\pi}{4} + arcsin\left(\frac{sin(\alpha)}{\eta_1}\right)\right) - \eta_2^2\right]^{-\frac{1}{2}}, \tag{10}$$

where:
α - angle of the incident beam,
η_1, η_2- refractive index of glass and water,
Λ - laser wavelength.

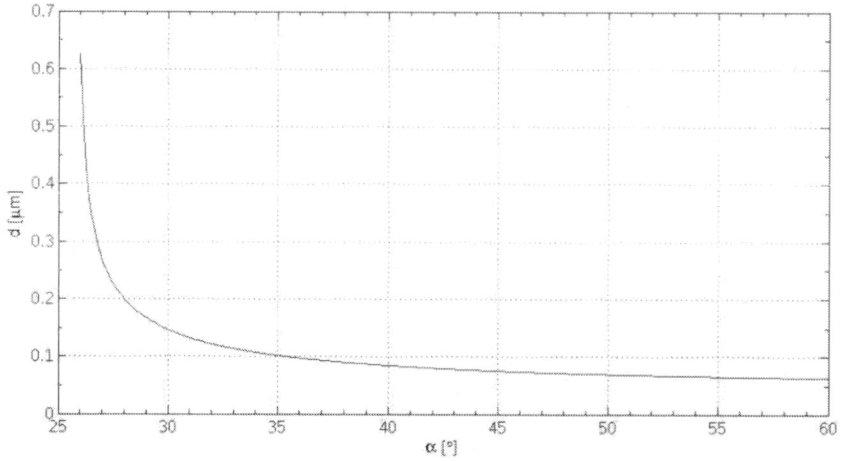

Figure 5. Penetration depth as a function of light incidence angle.

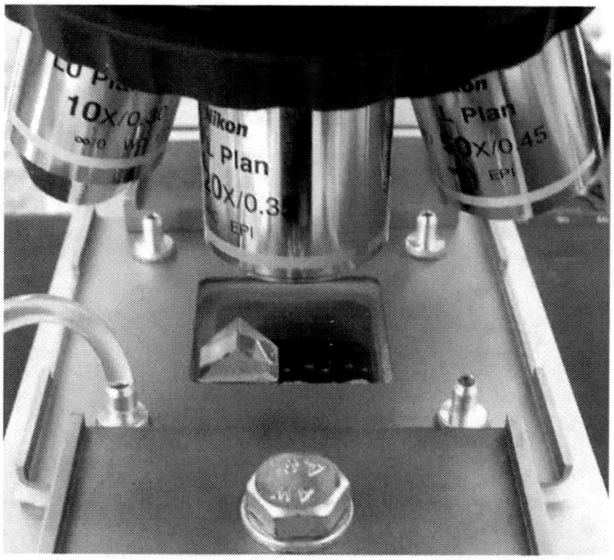

Figure 6. Microchannel under the microscope. Total internal reflections of the laser in the upper wall can be seen.

Fine adjustment of the laser illumination angle allowed us for a precise control of the penetration depth of the evanescent wave, which was set to be $d = 107 \pm 6nm$. This way only the particles within the penetration depth where illuminated strongly enough to be detected by the camera. The measurements were performed under an epi-fluorescent microscope (Nikon Eclipse $50i$), for particles located in the $107nm$ light sheet close to the wall and in the centre plane of the channel height (Fig. 6). Microscopic images were acquired using a high speed camera ($PCO1200HS$). The evanescent wave illumination was formed using $1kHz$ Nd:YAG pulsed laser (Ekspla $NL202/SH/TH$) with $532nm$ light wavelength. The temporary positions of the particles were evaluated from the long sequences of acquired microscopic images by means of especially developed filtration and image analysis MATLAB codes. Each sequence consists of 650 images. In order to reduce static noise, a mean image of a sequence was calculated and then subtracted from each frame. This way only moving particles were clearly visible. To find the exact positions of the particles, each point of a frame was correlated with a theoretical model of a particle image. Places where the correlation value was above 0.6 particle position was found. Since light intensity of a particle image has a two dimensional Gaussian profile, the theoretical model image (Fig. 7) is given by the following equation:

$$f(x,y) = exp\left(-\frac{x^2+y^2}{2\sigma^2}\right). \tag{11}$$

Figures 8 and 9 show a typical image before and after the applied filtration. In the second

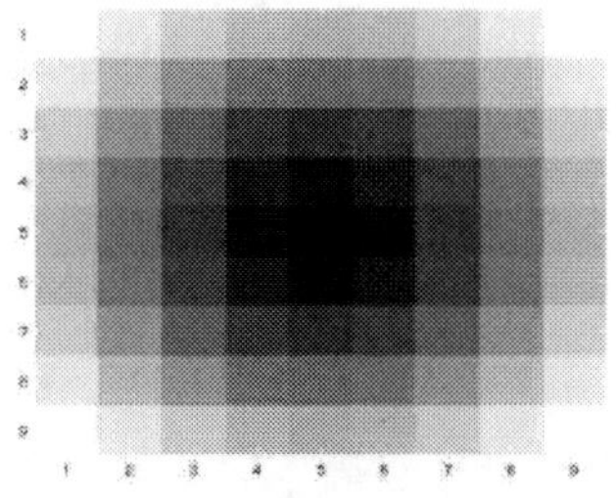

Figure 7. Computer model of a particle image.

image one may easy observe detected particles encircled for better visibility.

Once all of the images were analyzed, the identified particles were used to build individual particle trajectories. The concentration of particles was set so that the maximal displacement of a single particle between frames is much smaller than the mean distance between other particles. That way the algorithm for finding trajectories won't mix the particles and we minimize the possibility of producing false displacements.

Images where calibrated using a precise optical scale. For the lens of magnification 20x the pixel square size was $0,6\mu m$, and for the objective of magnification $60x$ the size was $0,25\mu m$. Figure 11 shows a typical trajectory of one of the particles. The evanescent wave illumination used had penetration depth of 107nm and for 300nm particles only those which center was closer than about 200nm from the wall could be visualized. With decreasing distance from the wall their images became brighter and larger. Precise evaluation of the size of particle spots on the images allowed us to find their apparent diameter, hence to evaluate their distance from the wall. The average distance of a particle center from the wall was found this way to be about $h = 170nm$.

The diffusion coefficient was calculated in the same manner as in the numerical simulation using equation (9). Since we only observe particle movement in the plane parallel to the channel wall, only the parallel diffusion coefficient can be obtained from the present experiment. Two sets of measurements were performed. Measurements far from the wall were performed by observing particles in the middle of the channel and illuminating them by a fluorescent lamp of the microscope. Measurements close to the wall were done using evanescent wave illumination. From the measurements performed in the middle of the microchannel height we obtained the diffusion constant in the bulk of the fluid. By doing measurements close to the wall we obtained the parallel diffusion constant for particles close

Figure 8. Image of particles close to the wall with a lot of static noise before filtration.

Figure 9. The same image after filtration and noise reduction, with only moving particles localized.

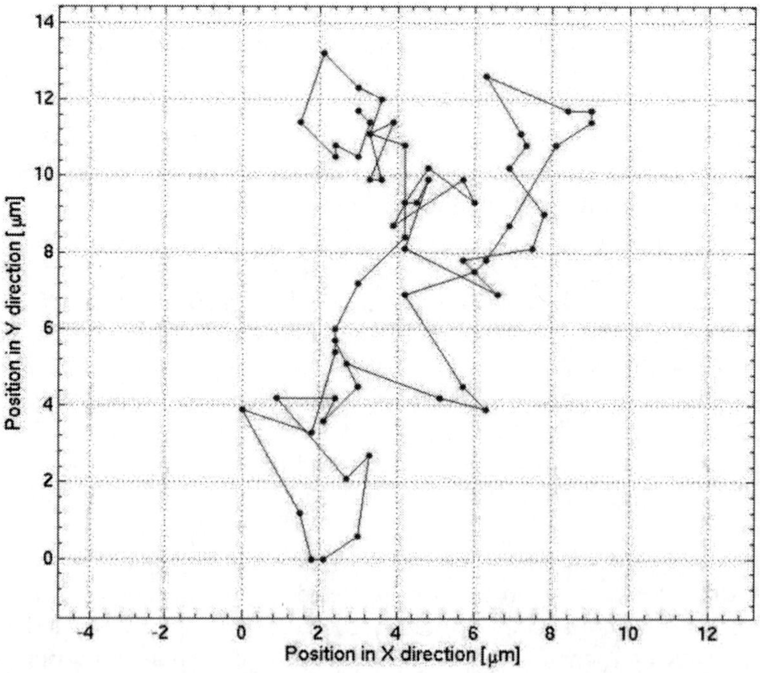

Figure 10. Typical trajectory of a Brownian motion. The particle was $300nm$ in diameter, time between points is $0.5s$.

Journal of Physics: Conference Series **392** (2012) 012014 doi:10.1088/1742-6596/392/1/012014

Table 1. Diffusion coefficients measured in the cavity centre and close to the wall.

In the bulk		Near the wall	
Measurement	$D_{\|\|} * 10^{-12} m^2/s$	Measurement	$D_{\|\|} * 10^{-12} m^2/s$
I	1.49	I	0.79
II	1.30	II	0.83
III	1.10	III	0.84
IV	1.33	IV	0.82
V	1.19	V	0.92
mean	1.28	mean	0.84

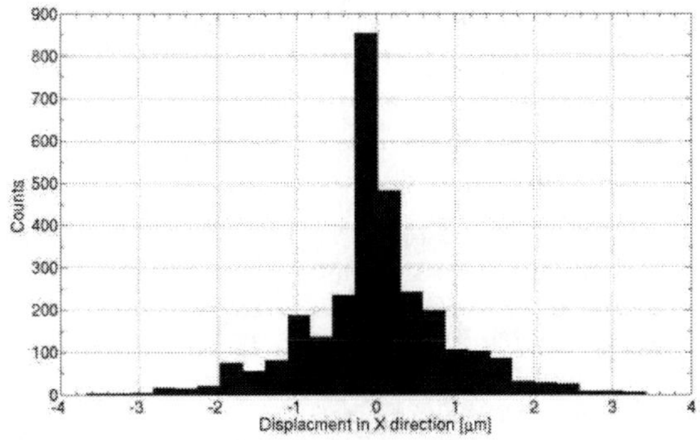

Figure 11. Histogram showing particle displacements parallel to the wall in the X direction; selected time step is 0.5s.

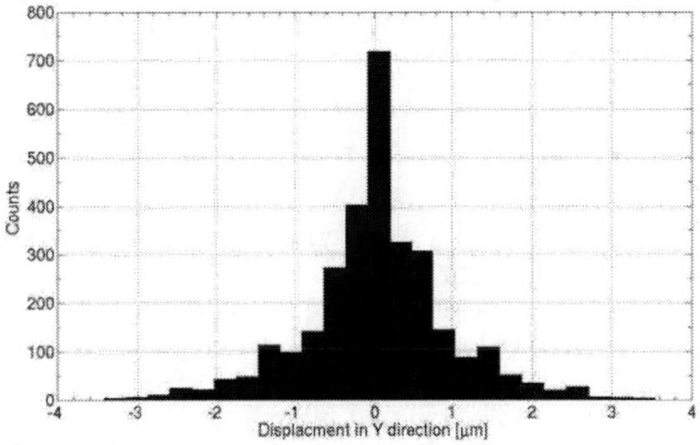

Figure 12. Histogram showing particle displacements parallel to the wall in the Y direction for the same experimental set.

to a solid surface. A total of 10 measurements sets where preformed, 5 close to the wall and 5 in the bulk. The results are displayed in the Table 1.

Figures 11 and 12 illustrate displacements for 300nm particles close to the wall measured in two perpendicular direction, both parallel to the wall. These two displacement components were used to evaluate the diffusion constant.

By calculating the mean diffusion coefficient and by knowing the particle distance form the wall we can estimate the slip length using formulations given by (5). Based on these measurements near wall diffusion coefficient is $\beta_{||} = 0.66$. For the average particle distance from the wall $h = 170nm$ the evaluated slip length is given as $\lambda = 300nm$.

This relatively large value still remains within spread of data reported in the literature ($5nm - 10\mu m$). Usually large slip length is associated with superhydrophobic surface properties or high surface roughness. In our case measured average roughness of the quartz glass used was below 1nm. Detailed study of the wall surface under AFM microscope excluded presence of gas pockets or nano-bubbles, which are often reported as slip triggering agents [3]. Also the measured wetting angle for the glass wall was in average equal 40° discarding effects of strong hydrophobic wall properties.

4. Conclusions

Our preliminary numerical simulations suggest absence or very small slip effect for the Brownian motion of nanoparticles. It is generally in agreement with reported MD simulations for the slip length found for shear flow in a microchannel. It was shown that substantial slip develops in nonwetting situations when the contact angle is larger than 90°, with slip lengths reaching 10 - 50 molecular sizes [12]. Nevertheless, it must be concluded that molecular simulations of Brownian motion performed for short time scales and constrained by dimensions of several hundreds nanometers may not properly represent long time self diffusion coefficient of the particle. This work is still in progress. Additional numerical test are necessary to elucidate how far the slip length obtained for Brownian motion relates to that calculated for shear flow.

Our preliminary experimental data for the slip length indicates relatively large value of about $0.3\mu m$. This value is quite large not only in contrast to the simulations but also compared with several experiments performed for hydrophilic surfaces. This may be due to the relatively large particles used, comparable to the distance from the wall and to the penetration depth of the evanescent wave. Since the camera used has a large noise, it was impossible to determine the exact particle distance h from the wall based on the particle image size and its intensity. The work is continued using smaller particles ($24nm$) and using a new intensified camera. We will be also taking into account modification of the wall properties to verify effect of hydrophobic and hydrophilic interactions.

References

[1] Holt J K, Park H G, Wang Y, Stadermann M, Artyukhin A B, Grigoropoulos C P, Noy A, Bakajin O 2006 *Science* **312**, 1034-1037
[2] Neto C, Evans D R, Bonaccurso E, Butt H-J, Craig V S J 2005 *Rep. Prog. Phys.* **68**, 2859-2897
[3] Lauga E, Brenner M P, Stone H A 2007 In *Handbook of Experimental Fluid Dynamics* (ed. J. Foss, C. Tropea, A. Yarin), chap. 17, 1219-1240, Springer
[4] Martini A, Roxin A, Snurr R Q, Wang Q, Lichter S 2008 *J. Fluid Mech.* **600**, 257-269
[5] Kalwarczyk T, Zieebacz N, Bielejewska A, Zaboklicka E, Koynov K, Szymanski J, Wilk A, Patkowski A, Gapinski J, Butt H-J, Holyst R 2011 *NanoLetters* **11**, 2157-2163
[6] Fakhri N, MacKintosh F C, Lounis B, Cognet L, Pasquali M 2010 *Science* **330**, 1804-1807
[7] Franosch T, Grimm M, Belushkin M, Mor F, Foffi G, Forro L, Jeney S 2011 *Nature* **478**, 85-88
[8] Lumma D, Best A, Gansen A, Feuillebois F, Radler J O, Vinogradova O I 2003 *Physical Review* E **67**, 056313
[9] Tretheway D C, Meinhart C D 2002 *Phys. Fluids* **14**, L9-L12
[10] Tretheway D C, Meinhart C D 2004 *Phys. Fluids* **16**, 1509-1515
[11] Joseph P, Tabeling P 2005 *Phys. Rev.* E **71**, 035303
[12] Cottin-Bizonne C, Cross B, Steinberger A, Charlaix E 2005 *Phys. Rev. Lett.* **94**, 056102
[13] Lauga E, Squires T M 2005 *Phys. Fluids* **17**, 103102

AUTHOR INDEX

Institute of Physics
Dirac House, Temple Back
Bristol BS1 6BE UK

CURRAN ASSOCIATES INC
proceedings
.com